D0919409

3 0700 10352 3253

Using Video: Interactive and Linear Designs

Techniques in
Training and Performance Development
Series

Using Video: Interactive and Linear Designs

Joseph W. Arwady and Diane M. Gayeski

Joseph W. Arwady
Series Developer and Editor

Educational Technology Publications
Englewood Cliffs, New Jersey 07632

Library of Congress Cataloging-in-Publication Data

Arwady, Joseph W.
Using video.

(Techniques in training and performance development series)
Includes index.
Bibliography: p.
1. Video recordings—Production and direction.
2. Interactive video. 3. Employees, Training of—Audio-visual aids. 4. Video tapes in education.
I. Gayeski, Diane M. (Diane Mary), 1953-
II. Title. III. Series.
PN1992.94.A79 1989 791.45'0232 88-31105
ISBN 0-87778-199-0

Printed in the United States of America.

Library of Congress Catalog Card Number:
88-31105.

International Standard Book Number:
0-87778-199-0.

First Printing: January, 1989.

Dedication

To our spouses, Meriette and David, who learned long ago to separate good video from bad. And to our children, Tom, Abigail and Evan, who are still learning to discriminate.

Acknowledgment

We are indebted to Brook Ashby, Director of Operations at Penfield Productions in Agawam, Massachusetts, a firm specializing in corporate training and promotion. Brook is a talented friend and associate who has generously contributed ideas and support material to our book.

Editor's Preface

There are video books about production that offer good, practical lessons in sound, lighting, graphics, editing, special effects and other technical areas. The exemplar in this category is Gerald Millerson's *The Technique of Television Production*. Other books are less technical in their orientation and cover a broader range of conceptual material. This second category is characteristically about managing video for particular audiences or settings. *Corporate and Instructional Video*, by my co-author, Diane Gayeski, is a good example of a nonbroadcast textbook that blends theory and practice. Still other video books report state-of-the-art surveys of the industry, its people, and the medium itself. These books take an even broader view of video, seeking to measure its effects, potential, and direction. Judith and Douglas Brush have compiled several surveys, known as the Brush Reports, that include their prognoses for the nonbroadcast industry.

Using Video: Interactive and Linear Designs doesn't fit in any of these categories. It champions *design* as the most important phase in any production and it argues that specific *design techniques* are the most potent vehicles for increasing a program's capacity to manipulate an audience. And isn't that what video is all about—*manipulating the audience*? The 40 techniques in this book, 22 interactive and 18 linear, are design tools around which entire programs can be built. Use them to make the audience think, to get a laugh, to change behavior, and to improve performance—all responses that signal the program *works*.

I can attest to the fact that something else in this book "is working," and that's my co-author, Diane Gayeski. It's been my good fortune to collaborate with someone so thoroughly competent and supportive. Gayeski is a video maven. She has 15 years of professional production experience to her credit and is a respected spokesperson for the non-broadcast industry. Her 1983 book, *Corporate and Instructional Video*, is about managing and designing non-broadcast programming, two functions she not only writes about, but also practices. I've seen several Gayeski productions and there is no mistaking her talent for balancing creative design with practical, results-focused programming.

Her firm, OmniCom Associates, counts Xerox, Marine Midland Bank, Carrier Corporation, and University of Helsinki among its clients who contracted for an initial project and have come back for more—and not necessarily more of the same. Where interactive video remains one of OmniCom's primary calling cards, they often recommend and produce linear video and computer-aided instruction as well and teach their clients to develop these media themselves. And while it's comforting for clients to know that Gayeski and her staff can work in different technologies, it's the quality of program design that makes the real difference.

Diane and David Williams, her partner in OmniCom, are also professors at Ithaca College. The Communication Program she helped to build at Ithaca is testament to her work with thousands of aspiring video professionals. And the job placements are of the quality and number befitting what many claim is the best undergraduate program in Communications, anywhere. No doubt Professor Gayeski has contributed to Ithaca's reputation, but so has she benefitted by "rubbing shoulders" with some of the nation's brightest and most enthusiastic students of video.

My own "video baptism" began the day I started work at Litton Business Systems. I had left academia for the business world, joining Litton as Manager of Training and Manpower Development. Chris Dunlap, Vice President of Manpower Development at the time and a veteran in the business of office systems training, invited me into his office for a first-day briefing. After a "how are you?" and "glad to have you on board," he dropped my first assignment: "In six weeks, purchase and install a 300-location video network *and* produce a program that introduces video with a 'bang.' And make sure it promotes the company's new product lines in computers, copiers, and calculators." Six weeks and nearly half a million dollars later, we had succeeded on all counts, but not without our share of thrills and chills.

First of all, once the word "hit the street" that a major video purchase was up for bid, every vendor in New York and New Jersey began calling and referring to me on a first-name basis. "Joe, baby, this is Irv. Here's the best I can do for you. What? Still too high? I'll call you right back." This went on for five weeks until we reached agreement with Panasonic in Japan for drop shipment to all locations on a single, specified date. Working though L. Matthew Miller, a Manhattan wholesaler, the "equipment side" of the deal went off without a hitch, Their man, Lon Mass, now with Sony Broadcast, had pity on me and engineered every detail.

Meanwhile on the production side, we decided to tape the morning session of a nationwide branch manager's meeting, edit during the next day, and deliver finished programs to 300 departing managers on day three. They would conveniently return to their branches where hungry sales reps and a "surprise" video equipment delivery would be waiting. Bringing the highlights of the company's most exciting branch manager's meeting to field sales and service reps would be the vehicle Litton used to launch three new product lines in 300 locations on the same day. And believe it or not, it worked!

Of course. there was the 22-hour all-night editing session at PolyCom's new Chicago production facility. I can hear my editor droning the words "Java. I need more java" during most of the night. Even Carmen Trombetta, PolyCom's executive producer, was surprised to see us there the next morning. We didn't look quite the same as we had the previous morning. "Disheveled" would be a good description.

And I'll never forget the multi-site, multi-vendor duplication frenzy we required in order to deliver those first tapes on time. I remember waiting outside the meeting room hoping the remaining 100 tapes would arrive before President Bob Kane sent those managers home. I kept murmuring, "Keep talking, Bob. Just a few more minutes." He did, the tapes arrived, and the video adventure began with a plus. Our reputation never suffered after that. More than 20 productions later, when I left Litton, video remained an effective and credible medium.

Since Litton, I've worked as a consultant on program design and scripting, with an occasional stint as producer/director. My current work on productivity and performance management is several steps removed from hands-on video, but the similarities far outweigh the differences. Any area of management is concerned with measurement and manipulation. The better our measures of the target audience and work environment, the more creative our designs and effective our manipulations. And the more likely the audience will understand, react, and accomplish.

Throughout this book, we've tried to keep our eye on the viewer, the end-user, who in a strange twist of priorities is a "given" in many corporate productions. Instead of being the target of considered analysis, instead of driving design decisions, the audience and their responses to programming often receive little more than passing comment. Once the focus shifts to production mechanics, it's full speed ahead until the program is "in the can."

We hope to temper that laissez-faire attitude about viewer response and behavior. We hope also to influence some of your thoughts about audience analysis and the ways people process video information. The interaction of programming, cognition and performance is complex, but nonetheless manageable. Especially in the world of work, clear objectives and thoughtful design can shape audience response and behavior.

Before you finish reading *Using Video: Interactive and Linear Designs,* you will see why we place such high value on script-writing and video design. Our 40 techniques will either start you on the road to better video or accelerate the trip you've already begun.

Joseph W. Arwady

Table of Contents

Dedication. v

Acknowledgment vii

Editor's Preface. ix

1. Introduction: Using Video: Interactive and
   Linear Designs. 3

2. **Linear Design Techniques** 7
   Eighteenth Hole Technique. 15
   Audio Lead Technique 21
   Off-Screen Audio Montage Technique. 26
   Using Music Technique 31
   Spokesperson Technique. 33
   Out-of-Character Commentary Technique. 39
   Subjective Viewpoint Technique. 43
   Character Tracking Technique. 49
   Alter Ego Technique 54
   Parallel Scenes (Action) Technique 59
   Quick Visual Inserts Technique. 65
   Flashforward Technique 72
   Graphics and Animation Techniques. 77
   Cue Fading Technique. 80
   Repetition Technique 82
   Deliberate Degradation of Video and/or
      Audio Technique 88
   Deliberate Overstatement (Exaggeration)
      Technique 90
   Dramatic Irony Technique 97

3. **Interactive Design Techniques** 101
Pause Technique 108
Controlled Playback Technique 111
Introductory Menu Technique 113
Intelligent Manual Technique 117
Visual Data Base Technique 121
Pretest Technique 124
Selecting Relevant Examples Technique 127
Vicarious Travel Technique 131
Third Person "Directed" Simulation Technique 135
First Person Simulation Technique 137
Simulation with Feedback Technique 140
Gaming Technique 142
Tutorial Technique 145
Opinion Question Technique 148
Auditory-Visual Test Technique 150
Response Peripheral Technique 153
User Comment File Technique 158
Learning Style Diagnosis and Branching Technique 160
Videotape Responses Technique 163
Consensus Technique 165
Hidden Programming Technique 168
Intelligent "Recursive" Programming Technique 170

Bibliography 173

Index 175

Using Video: Interactive and Linear Designs

1

Introduction: Using Video: Interactive and Linear Designs

Video . . . it's such a familiar medium—yet it's still one of the most misunderstood. It looks so easy—but we know there are a thousand complicated, technical details that need to be mastered. It can be so captivating—yet we've all been bored to tears by misuses of the technology. Programs can be produced for $10.00 or over $100,000. You can use it to demonstrate "how to" or to record a panel of experts, or to produce a network-style documentary, or a game show take-off, or a company news update. And then there's interactive video, with its computers, videodiscs, diskettes, touchscreens, and other so-called mysterious and complex hardware.

Video is arguably the richest of all media, given its capacity to display varieties of auditory, visual, and emotional stimuli (and in the case of interactive video, to respond directly to the viewer). Despite this incredible array of production options, the fundamental purpose of video is not to dazzle the audience, but to *manipulate* the viewer. Video is an intimate medium, best suited for small group or independent viewing. In the non-broadcast environment, what happens *after* video is usually the basis for judging whether a program is effective in manipulating the viewer, i.e., stimulating the desired response. Video doesn't have to be pretty to work, but it must be cleverly designed.

Clever design is logical, grabbing the audience up front and transporting it through a viewing encounter that demands attention, anticipation, and response. These are mental activities that signal the viewer's involvement with the program. The most expensive production in the world is nothing more than the most expensive production in the world, if it fails to elicit active mental processing from the viewer. This is true for interactive as well as linear programming. Some interactive programs are so poorly structured that the viewer (user) is quickly bored and uninvolved, despite branching and intermittent physical contact with hardware. To make good decisions about design, to understand program structure, it is necessary to segment video's capabilities into manageable chunks. That's exactly what we've done in this book.

In *Using Video: Interactive and Linear Designs*, we've selected 40 video techniques, certainly not the universe, but a sampling of those we consider most useful (i.e., manipulative). All 40 have stood the test of repeated application in non-broadcast settings, including those occasions in business, industry, government, and education where we've used them ourselves. Each technique is a *design vehicle* that you can incorporate into programming to achieve specific objectives.

Professionals in the instruction and information business will find some of the techniques familiar; after all, most of us spend the equivalent of one work-day per week watching broadcast television. And video is a close runner-up to stand-up lectures in terms of its prevalence as a training method in today's organizations. Perhaps you're already using video in training; according to recent surveys, almost 90% of companies with more than 100 employees are, too. That's about 8,500 organizations, so you're in good company. About half of large organizations (over 10,000 employees) have at least experimented with interactive video, but even this technology is not limited to those with large budgets. Many of the most effective programs have been produced by the smallest organizations with minimal resources. Although many organizations are as yet using only "off-the-shelf" programs, an increasing percentage are producing their own. This custom programming may be generated in-house, or by vendors, or by a mix of both resources. But even when productions are contracted out, an organization still has to supply expertise in terms of content and analyses of the audience and settings in which a program will be used.

Unfortunately, despite its widespread application, video is often not used to its full capabilities. Much emphasis has been placed on production equipment and technology–such factors as crispness of the picture, elaborate special effects, and computer graphics. Although these considerations and devices can make programs look "slicker," they usually have no effect whatsoever on what is learned. For those of us in the performance and training business, this means that we've got to concentrate on other factors, namely the design of the program and the *specific scripting techniques* that can be applied to instruct, motivate, and even entertain.

You will find techniques here that remind you of a program or two you've seen. Other techniques, especially the interactive video ones, may be less familiar. But when you've got a training or information problem and it seems like video could be the solution–or when you're staring at a blank sheet of paper that should be a script or a request for a proposal from a production house–this book will help you get started. We'll identify each technique or program format, explain what it is and why it works, give you some hints and cautions about its implementation, and present some examples of how organizations have used it successfully in instructional and informational applications. We're betting that our treatment of these techniques will provide you with new insights into *how to manipulate video programming.*

A number of factors contribute to the success of video and interactive video as training and information vehicles within an organization. Before you launch into video, here are some questions you might consider:

1. **Is the topic appropriate for video?** Can it use sound, motion, and color, or is it better suited to a handout, job aid, or slide show? If something can't be visualized or dramatized, other media are probably cheaper, more appropriate, and easier to use. Although video can be quite entertaining, used inappropriately, it can be deadly. For instance, teaching sales representatives to fill out travel expense forms could be done on video, but little can be done with the topic except to show close-ups of the form and perhaps examples of it being filled out. Perhaps a job aid consisting of some explanations of the terminology and procedures–or even better–a re-design of the form which would make costly training unnecessary would be a better solution.

2. **How will the program be used?** Is it meant to be used in a classroom setting, or by individuals—or both? Is it part of a larger package, for instance, a workbook/video presentation, or is it meant to stand alone?

3. **Do you have a delivery system that's convenient, easy to operate, readily accessible, and in appropriate locations?** If your audience has no equipment, or it's never hooked up, or there's no place to watch a program, the best videotape or videodisc will never be seen. If programs are to be used in dispersed settings, are there individuals at each location familiar with and responsible for the equipment? This is especially crucial in interactive video, where the hardware systems are more complex, and one loose wire can cause the entire system to malfunction.

4. **Is there organizational support for video programming?** If trainees feel threatened by the medium, you might require that they use it, but they'll probably do so poorly. If your audience sees it as an expensive frill rather than a cost-effective tool, you'll only reap bad feelings. And if management likes the "idea" of video, but harbors low expectations for results, support will erode quickly.

5. **Is there a feedback mechanism available?** As we all know, communication by definition is two-way. Because video is often used to take the place of classroom presentations which involve travel and scheduling constraints, there is frequently little opportunity to get direct or even indirect evaluation of programs. Without this, there is obviously no means to assess the effectiveness of the programs and to improve future productions.

2

Linear Design Techniques

Video! Just the sound of the word carries a measure of excitement . . . at least for a while. That is until you've been through twenty or thirty productions. Somewhere along the way, the glamour wears off and reality sets in. Scripting, production scheduling, programming, casting, computer interfacing, distribution, audience analysis, measuring impact—it's not easy! Video, despite its vast potential and innumerable possibilities, remains very vulnerable. It can be poorly executed, sloppily managed and, most seriously, ineffective.

The linear portion of this book is not designed to address production issues per se. Nor is it intended to solve most programming problems. What it does is present a collection of techniques that can influence the viewer to participate actively in the interpretation of program information.

Talking heads, role plays and magazine formats don't do that. They have their purposes, but fall short of *stretching the mind*, of taking the viewer to a response point where he or she unavoidably plays a part in determining the program's impact.

This book's primary audience is the corporate or industrial professional who is interested in purposefully selecting and utilizing specific video techniques to heighten viewer involvement and influence viewer response. The behind-the-scenes technical wizard or the uninvolved "just keep me informed" client representative are not the book's target audience, and the techniques are not intended for them. Each technique is included explicitly because it has been used successfully to establish a level of interaction where viewer cognition is influenced in ways that can help shape subsequent behavior and performance. In the right hands, the 18 linear techniques in the first part of the book are potent tools for stimulating imagination, conveying information, focusing attention, emphasizing important ideas, communicating attitudes and emotions, establishing relationships, modeling behavior, and personalizing screen action.

Knowing Your Audience

Perhaps the highest priority in conceptualizing a program is to *know the audience*. Here, we refer to much more than knowing the audience as sales reps, or computer sales reps, or even computer sales reps in Los Angeles with a six-month opportunity window. We need to gather pertinent information about an audience, especially as *performers* in the work environment, in order to improve program content, technique and impact. Exactly what is "pertinent information?"

To begin with, it should be *task specific*. It's nice to know that audience members are civil engineers, but unless that information helps in understanding them as *viewer-performers*, it is of little value. It would be better to know that an audience is comprised of two hundred civil engineers who are working on designs of military access roads, which must pass three Defense Department reviews to insure continuation of funding, and so on. The more you know about specific job functions that viewers must perform, the better equipped you are to design programming that triggers viewer recognition and response.

A second facet of good audience analysis is acquiring an *intimate knowledge* of the prospective viewers. This is really a second phase or byproduct of descriptive data collection which grows from increasing familiarity with the viewers and their work environment(s). It's the kind of unwritten or tacit understanding that says "that is what makes these people tick . . . this is why they'll respond to the program." Intricate knowledge doesn't come from outside producers or script writers. Nor does it come from insiders with isolated staff roles. It does come from content experts–the employees or former employees who have firsthand experience in the work environment.

A third consideration in knowing the audience is the broad area of *viewer aptitudes*. When an individual sits down to watch a video program, he or she comes equipped with an incredibly complex array of mental skills and information processing preferences. These factors or *aptitudes* work in combination to influence how video messages are perceived, recoded, and stored in the viewer's brain. Some examples of aptitudes which shape the viewing experience are verbal skills, cultural biases, job knowledge, sense of humor, attention to detail, personality traits, and content familiarity.

While no one would suggest conducting a full-blown psychological analysis of prospective viewers, in the case of linear video it is practical to define the audience in terms of *shared aptitudes*. For example, "these are professional pilots with low tolerance for ambiguity, high self-regard and poor interpersonal skills." With interactive video, we go a step further and attempt to "fit" the program to individual viewers. This is accomplished by using a knowledge of the audience to produce *branches* or program options for different types of viewers. More about branching and its various levels later in the interactive section of the book.

In any case, your objective in knowing the audience is to structure the program to accommodate audience aptitudes and task requirements. When this is accomplished, the program is almost always distinguished by its freshness, its cleverness, and most importantly, by the impact it has on the audience.

Linear Techniques: Scriptwriting Tools

As with any technology, there are certain fundamental practices that distinguish a winning production from routine programming. In our view, none is as important or valuable as scriptwriting.

The video script is a roadmap for the entire production. It is synonymous with *program design*.

A good script *anticipates* what *should* happen in production and post-production.

A good script "sees" talent before they are cast and contours the program to maximize their on-camera presence.

A good script understands the individual viewer and group audience as consumers and manipulates program flow and perceptual cues to elicit emotion, involvement and reaction.

The script integrates sight and sound in a rich, sometimes complex relationship that leverages the medium's versatility to engage the audience. It builds *options* into the program, permitting the director to try several approaches, several shots, one of which will work best in the final presentation. And the good script is written for the editor, who will reassemble an entire production's worth of video segments and sounds into a finished product that is consistent with the program's design.

A good director shows his respect for a good script by shooting it, not rewriting it. He doesn't have to. With few exceptions, there is little reason to alter a shot, change a location or rewrite a line because again, a good script anticipates the the decisions that need to be made during a production. While talented production crews are impressive in improvisation, they much prefer contributing to an organized and well-managed production.

The 18 Techniques

The linear techniques included in this book are excellent scriptwriting tools. Anyone of them can draw attention to pertinent information, rehearse key ideas and involve the audience in *active viewing*. Some techniques are so dominant that an entire story line can be written around them. It is difficult to organize the 18 techniques in discrete categories because each offers the scriptwriter and director more than one way to improve a production. For the sake of discussion, however, we can describe the techniques in terms of their primary functions: (1) using *audio* to draw audience attention; (2) using *visuals* to alter audience perspective; (3) using *talent* to involve the audience in character development; and (4) using *humor* to get the audience to laugh and, at the same time, think and learn.

Audio

We often hear people refer to video as a visual medium. It's not. Video is a combination of audio and video signals, which are arranged in some meaningful way to inform, entertain, persuade, motivate or indoctrinate.

Audiences have difficulty remembering much of anything about most video programming, but are particularly apt to draw a blank when it comes to audio. Often, untrained scriptwriters will write the audio portion of a script first, treating it as if they are writing prose. Video is not a "short story" and this "written language" approach ignores the medium's encyclopedia of visual expressions. By ignoring the medium's inherent visual impact, an "audio only" scriptwriter loses sight of where audio fits in the entire production and how it will sound to a "viewing audience."

Other programs evolve with a dominant visual element, leaving the audio to be "tacked on" later once the locations, sets and camera shots are blocked out. This approach promises that the audience will consider the audio as an afterthought as well.

Good scriptwriting aims to meet program objectives, specifically those that relate to viewer behavior. For this reason, audio must interact with visuals to produce a more lasting effect than either can produce independently. This is a paramount rule in video design and the reason for The Eighteenth Hole Voice, Audio Lead, Off-Screen Audio Montage and Using Music techniques.

Talent

No device offers the scriptwriter or director more potential ways to reach an audience than does the talent. The human torso, face and voice bring with them an endless range of attitude, intonation, gesticulation and expression. A knowing glance, an infectious laugh, a crazed reaction, a personal reference—any of these can attract the viewer, hold his attention and set-up a one-to-one exchange with a character.

The five "talent" techniques selected for our book are potent in both their capacity to carry an entire story line and engage the audience in a "relationship" with key characters. All five afford the viewer a special status in sharing insights, discovering subtleties, and generally going beyond the third-party viewing experience so common to most programming. We recommend frequent experimentation with the Omniscient Spokesperson, Out-of-Character Commentary, Subjective Viewpoint, Character Tracking and Alter Ego techniques.

Visual Variations

If we consider a chronological sequence of events as "normal" visual progression, anything that deviates from the expected becomes novel. Granted, it's not enough simply to be different; you need to tie innovative visual formats to desirable audience response.

This is an important point and one that is often lost in the "deadline and dollar" world of corporate video. Even though parallel action or flashbacks enhance program form, they and other visual devices must also connect screen action to audience aptitudes and/or work performance. Good drama about the dangers of drinking and driving is nothing more than good drama if viewers fail to see how it pertains to *their* drinking and driving. For non-broadcast video to work, visual devices must drive programming into the audience's collective gut, where it says something more than it otherwise would about their personal reaction during and after the program. With rare exception, it's not enough simply to entertain.

The visual techniques included here can serve various practical functions, but are generally employed in predictable ways. Graphics and Animation, Cue Fading and Repetition are primarily used to reduce confusion, simplify complexity and focus audience attention on key information. Deliberate Degradation can also focus attention on those elements in a scene where production values are maintained. Or, it can create interest by dramatizing a particular time, place or character perspective, e.g., dream scenes that are shot slightly out-of-focus. Parallel Scenes is an umbrella term for a variety of visual devices that tie story lines together. Quick Visual Inserts and Flashforwards show relationships and highlight visual elements, alerting the audience to pay special attention.

Humor

One of the best ways to liven up a video session is to build humor into the program. Especially when the humor is an integral part of a scene, it can humanize the approach and add credibility to characterizations and content. In fact, the closer the connection between humor and realism, the funnier the scene. There must be "pain" in the portrayal for the audience to recognize the authenticity of the circumstances and see it as part of their world.

It is difficult to produce a program that is predictably funny with different audiences. There is always the risk that the context for the original script could change and what *should* have been funny is actually tragic. Beyond the conditions that govern the production, every audience views the program from a different perspective, as does each individual viewer. Make no mistake, producing humor is a demanding and difficult proposition.

In the non-broadcast video arena, two techniques in particular seem to be worth the risk. Both Dramatic Irony and Deliberate Overstatement (Exaggeration) permit the production team to manipulate real-world situations while actually improving the prospects for learning and performance. With the focus on information, training, and motivation, humor that dramatizes the relationship between behavior and consequence is worth its weight in gold.

Commercials and Linear Program Design

"We interrupt this introduction . . ." for a closing observation about commercial television. Here is a world very different from that of corporate/industrial video. Or is it?

Yes, if you consider that the objective of broacast television is to provide entertainment "at the margin,"* where audience ratings and time slots are sold to advertisers. No, if you look beyond the network shows to the real force in commercial video—the commercials! It's the commercials themselves that are affiliated with good design and the 18 linear techniques in this book.

Here, the programming is more difficult to interpret. It is not as "easy" to watch. The innuendo, imaging and symbolism is far more sophisticated. Most importantly, because greater *mental effort* is required to fully understand its meaning, the commercial is often *remembered* and *applied* when television watchers become product consumers. So while the actual programs are designed to entertain at margin, the commercials are written and produced at the highest creative level possible. Any producer who risks some-

*"At the margin" is an economic concept. It denotes the interaction of audience ratings, programming costs, and advertising rates, where profitability is maximized. A successful program controls costs, achieves competitive ratings and attracts healthy advertising revenues. Quality may only be "at the margin," but nothing more is required.

thing less is being pennywise and pound foolish. The 20-second spot is not the end-product; it is a vehicle used to accomplish another objective—to convince viewers to buy products.

The fact is that the real show in the world of commercial television are the commercials. The programs are squeezed in between.

These ten to sixty second "mini-programs" are generally well executed in order to (1) attract attention, (2) hold interest, (3) position the product, (4) trigger emotional involvement, (5) encourage buying behaviors, and (6) leave a lasting impression.

In writing the book, we found it interesting that the linear techniques we selected were prevalent in commercials but less often found in network programming. There's an analogy of sorts between our techniques and those broadcast spots. Corporate/industrial video has a distinct purpose and clear objectives. It must improve the viewer/employee's capacity to perform on the job. Similarly, commercials must sell products. If consumers don't perform (buy), advertisers won't advertise. If corporate/industrial programming doesn't communicate a message effectively, employees won't respond in the workplace and decision-makers will seek alternative means of communication. Both of these vehicles, the winning commercial and the effective industrial video, are task-specific. Both are designed to influence viewer reaction and behavior. And both make liberal use of the eighteen linear techniques that follow.

EIGHTEENTH HOLE VOICE TECHNIQUE

Purpose

To direct attention, create expectations, assess conditions or measure competence through the introduction of off-screen commentary.

Description

Delivered from a third-party perspective, the eighteenth hole voice is actually another name for a voice-over technique that anticipates, explains and evaluates behavior. Like the announcer off the edge of the eighteenth green during a televised golf tournament, the eighteenth hole voice is often hushed and dramatic. Used with a close-up shot, this "close-up" sound creates a feeling of intimacy and suspense.

Sometimes, the audience benefits from meeting the face with the voice early-on in the program. Brief introductory remarks by a moderator can provide a reference point for subsequent use of a voice-over. Another option is to introduce a character at the start of a program who establishes himself/herself dramatically as a knowledgeable and trusted authority figure. Removed visibly for most of the program, except perhaps for a late appearance to maintain continuity, the authority figure communicates periodically with the audience via voice-over.

It is also possible to introduce a voice without a visual reference. An initial observation from an unknown voice may startle the audience but at the same time presents an opportunity for a personal introduction. Imagine a shot of an interviewee sitting nervously in an employment office waiting for the interviewer to hang up the telephone and begin questioning. The eighteenth hole voice is first heard remarking:

> NARRATOR: (Off-camera female) "Look at his hands, sweating, cold, and clammy. If he rubs them together any harder, his skin is going to flake off (BRIEF PAUSE AS AUDIENCE SEES CS OF HANDS). Oh, my name is Giselle. I sort of follow Ted around to see how he's doing. You know, he's a real study in what not to do during an interview. If you don't believe me, just watch this."

In this example, the voice establishes an identity and role on its own merit. An animated and relaxed tone is conversational; it implies familiarity. This is someone who the audience enjoys listening to and would like to see because she sounds like someone they already know.

Whenever the eighteenth hole voice is used, it should be structured to attract attention to visual activity. The best time to introduce the voice is during a brief lull, just prior to key action. It takes longer to interpret audio information than video information, and screen action must be controlled to allow the audience time to concentrate on what is being said. The voice draws attention to a process and its outcome, while sharing an observer's point of view with the audience. Suspense is heightened because a character is just about to do something or an event is about to take place. Integrating the eighteenth hole voice requires the deemphasis or elimination of other audio sources and crisp transitions between what the voice describes and what the audience sees on the screen.

Often, the eighteenth hole voice is heard just prior to and during the developing sequence of events. It can be used to provide a descriptive, running account of key action as it takes place. Tracking the effects of important activity phases or steps in a procedure can reinforce recommended behaviors and relate them to rewards. Pared down narration and short phrases work especially well because they help the viewer focus attention and organize relationships. Even when it is used sparingly, the eighteenth hole voice can be an effective "teaching voice" in performance-oriented programs.

During a behavior modeling sequence, the technique offers the viewer an opportunity to evaluate two or three performance options. These can involve obvious procedural requirements or less common circumstances and more complex decision-making routines. As an example, consider a program designed to train bank tellers on security procedures during hold-up attempts. The eighteenth hold voice is heard to say:

> SPOKESMAN: (Off-camera, hushed voice) "The robber has entered the bank. Keep your eye on Kevin, the male teller, to see what he does when he realizes a weapon has been drawn. His objective is to maneuver the robber out of the bank as quickly as possible, reducing any risk of injury to employees or customers."

Obviously, in order to accomplish this two-part objective, Kevin must be prepared to execute a range of behaviors. His performance is dependent, in large part, on circumstances that evolve during the hold-up. But the eighteenth hole voice isn't used to enumerate all of Kevin's options, just as it isn't used to discuss the how-to's of putting during a televised golf tournament.

At key junctures in a modeling segment, the voice can build anticipation and suggest a mind-set for an imminent event or set of behaviors. The off-screen hushed commentary works its magic in small doses; too much detail or explanation kills the suspense and evolves rapidly into a dull, run-of-the-mill voice-over.

As the script develops and Kevin is required to make critical choices, the voice anticipates and even contributes to the development of these dramatic peaks. Suppose, as he is complying with the bank robber's request for money, Kevin intends to follow recommended procedures and press a floor-mounted alarm. The eighteenth hole voice, anticipating this action, continues:

> SPOKESMAN: (Off-camera, hushed voice) "Kevin seems to have the situation under control. He is in position to press the floor alarm. The key here is to move carefully and continue filling the bag as quickly as possible. So far, no one has been hurt."

The audience has a picture of what Kevin must do in order to activate the alarm. The bank tellers viewing the program are not casual observers; they are intent on observing how Kevin performs under such high risk conditions. The stage is set. The eighteenth hole voice has been used to raise anxiety levels, elicit an emotional response and bring the effects of prior training to bear on a dramatized version of a situation that every teller must be prepared to confront.

From this point, the script could develop in several ways. The robber could force Kevin away from the floor alarm and threaten to shoot at the next person who moves. Or, he could be oblivious to Kevin's successful effort to activate the alarm.

No matter how the script develops, the eighteenth-hole-voice can have a disruptive effect if used too often. Reserved for dramatic peaks when two or three contributing factors can be emphasized, the eighteenth hole voice filters out incidental elements and helps the audience focus on essentials. It is not the equivalent of a voice-over description by a moderator. It is not the source of a step-by-step accounting of events. The eighteenth-hole-voice is intended for situations where heightened suspense and anticipation can draw the audience in close for a concentrated view of key action.

Example

Bay State Gas used the Eighteenth Hole Voice Technique in a local commercial. The objective was to persuade users of oil heat to contract with Bay State for purchase or rental of an oil-to-gas conversion. The 60-second spot is very effective in (1) establishing a feeling of suspense and high expectation, (2) taking a rather ordinary event and lending it a degree of importance, and (3) adding an additional production device to what would otherwise be a straight talking head format. These are the same three functions served by the Eighteenth Hole Voice technique when it is used in corporate or industrial video. In either case, the technique combines verbal cueing with corresponding visual action, blocks out peripheral information and draws the audience to a single area of interest, and relies on the third-party, hushed-voice observer as a communication vehicle.

The Bay State Gas announcement opens with an establishing shot of a Victorian home on a quaint New England street, complete with sunshine and chirping birds. Enter the Bay State Gas service van and the third party spokesman who is carrying a large stop watch. Figure 1 captures this opening scene and includes the spokesman's on-the-spot, hushed remarks.

Figure 1. Eighteenth Hole Voice used in Bay State Gas commercial announcement courtesy Penfield Productions, Ltd. in Agawam, Massachusetts.

> Spokesman: "Don's about to convert this home from oil to gas heat. It will take him about 180 minutes on the average, but let's time him. (PRESS STOP WATCH.) That's a rental gas conversion burner he's about to install that will cost that family only $5.95 a month rental." (WIPE TO DENOTE PASSAGE OF TIME)

The spot finishes with Don emerging from the house beating the 180-minute average and the spokesman questions him about his performance. Much like the golfer who has just sunk a long putt, Don enters into a brief interview where he talks about Bay State Gas equipment, quality and easy installation. Basically, he credits the Bay State system and equipment for his success. The spokesman segues into a final remark about the conversion offer and how to obtain additional information.

There is a touch of humor in this example, as an ordinary serviceman is elevated to "sports hero" via the Eighteenth Hole

Voice technique. This aspect of the technique can be manipulated in any number of situations where the content is low risk and non-technical. In those cases where humor is advisable, the spokesperson can interpret the script, exaggerating the importance of the event and treating the performer as bigger than life.

AUDIO LEAD TECHNIQUE

Purpose

To prompt the viewer to anticipate visual content just before it appears on the screen by introducing a portion of the audio track from the next scene before making the customary visual transition.

Description

It is often said that video is a visual medium. While it's easy to understand the emphasis on the visual element, the statement is inaccurate. Actually, video programming is made up of sequences of visual *and* auditory signals which are organized in particular ways for different purposes.

Viewers expect that certain types of auditory and visual cues will combine to produce meaningful program content. In other words, what the audience hears is supposed to be synchronized with what the audience sees in ways that make sense. Any deviation from this pattern is likely to appear incongruous. Incongruity, however, when purposefully integrated into a program, can gain or hold attention and stimulate thinking (i.e., imagination).

The audio lead does prompt the viewer to make sense of incongruity and is also useful aesthetically in building dramatic value. The viewer's first response to an audio lead is "What is it?" followed immediately by "What does it mean?" Because it is something of a "brainteaser," the audio lead is best used in programs where scene-to-scene transitions are designed to be gradual and mildly perplexing.

The technique is simple to execute. Assume that an audio lead is used to move from Scene 1 to Scene 2. During scripting, transitions are written to accommodate the juxtaposition of audio information. The tail end of Scene 1 is scripted without audio to permit an audio dub or fade from the beginning of Scene 2. (See Figure 2.) If the audio lead consists of voice, it should be the length of an average sentence, say 10-15 words. Sound effects or music dubs need to be estimated for length, using the "average sentence" rule as a guide.

In general, keep it short. Viewers recognize the disparity between sight and sound almost immediately. They will try to make sense of the apparent incongruity. The audio lead need not be any longer than the time it takes to present a single thought. If the lead is too lengthy, the mismatch between auditory and visual cues ceases to be a source of mystery and heightened drama, appearing instead to be purposeless and poorly executed.

During post-production, a simple edit matches a few seconds of audio from Scene 2 with the final seconds of video in Scene 1. While the script normally provides direction on the location and length of the audio lead, there will always be some leeway in selecting final edit points that "feel right." The audio lead must establish itself quickly to gain attention and steadily overpower Scene 1 visuals in order to segue smoothly into Scene 2. Pre-edits are useful in locating in and out points that are sensitive to the visual flow in both scenes.

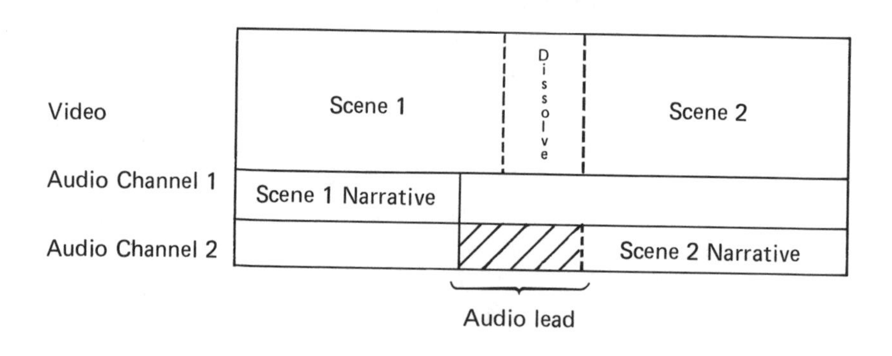

Figure 2. Graphic Depiction of Audio Lead.

Despite its brevity, the audio lead can trigger a high level of viewer involvement. It is not and does not appear as a separate audio device, produced independently, and inserted into the program. Rather, it is the audio portion of an existing scene that is excerpted and laid over part of the preceding scene. Keep in mind that while it "leads" the audience into the next scene, the visual part of the trailing scene is "sacrificed" in order to achieve the effect. The viewer hears part of the upcoming scene before seeing complementary visuals, so there is time for the viewer to take an active part in thinking about both the premature audio and the "missing" video.

When should audio leads be used? First, since time is required to involve the audience in the technique, audio leads should be used *between* action segments where the pace of the program is slowed. Audio leads lend themselves to between-scene transitions that are smooth and gradual and are not useful for moving from shot to shot *within* scenes like straight cuts.

Second, audio leads are used most often to transition ahead in time. The flow of most programs is chronological and the lead by itself does not prepare the viewer for a shift back in time. Additional cueing in the tail-end of the outgoing scene can position the audio lead to take the viewer back in time, although the video lead (flashback) is better suited to this effect.

Third, when quiet and loud scenes are juxtaposed, audio leads can be used to jar or excite the viewer, and when used with the scenes reversed, to effectively relax the viewer. It is the premature positioning of the audio lead as the bridge between scenes with different intensity levels that accounts for its ability to shock or sooth.

Finally, audio leads can be used to prompt the viewer to begin structuring a resolution to a problem or question. Since they do create a moment's confusion by presenting only the audio portion of a message, the audience has an opportunity to perform a few seconds of "mental editing" in anticipation of more complete information.

Examples

The audio lead can be used as a transitional device in programs where there is little continuity between segments. In these situations, the audio lead functions almost as a gimmick. It is not used to strengthen the relationship between the two scenes being bridged. It is simply a device that helps the viewer adjust to a change in subject matter.

For example, Paul Nemiroff (Paul Nemiroff Productions, Inc.) used the Audio Lead Technique in a highly creative and award-winning production for Sperry Corporation. The program "11:38: 24," was designed to familiarize Sperry employees worldwide with the wide range of products manufactured and distributed by its various divisions. The program was shot in documentary format, utilizing limited off-screen narration. Program segments are unrelated in terms of specific content, except that they all feature Sperry products and the ways they benefit mankind.

Where audio leads are used, they consist of peculiar voice and sound effects. Juxtaposed against incidental, ambient sound and over static visuals, the audio leads capture attention because they deviate from the typical pattern of synchronous visual and auditory cues. Without making an attempt to tie segments together substantively, they "surprise" the viewer and signal the start of new vignettes.

One product sequence features air traffic controllers relying on Sperry aircraft tracking systems. Near the end of the sequence, over ambient control tower sounds and a shot of blinking control panel lights, the audience hears, "That's it. Look up here. Everything looks fine." As the delayed visuals reveal, the next Sperry product is an electronic sensory implant used to rehabilitate patients with sight (eye) damage. The quiet, soothing voice is not that of an air traffic controller, but of an eye surgeon.

In a second example, a major chemical firm distributes a monthly news bulletin to its branch locations. Frequently, the format calls for the host moderator to provide background on a lead story, with testimonials and documentary footage used as follow-up and support. Audio leads are used to make the transition from the background to the "proof" stage of the story. Here is an excerpt from a typical program:

Video	**Audio**
MS technician working in lab	Moderator (VO): . . . . Then, the residual is tested for impurity levels that cannot exceed three parts per one million mg.
CS test area. Show technician's hands on glass tank.	Moderator (VO): What does exhaustive testing in sterilized glass tanks like these mean to your customer? You'd be surprised.

ECU test tank in motion	Tannenbaum (AUDIO LEAD): Quality control has always been the name of the game in chemical manufacturing.
MS Bob Tannenbaum, Customer. (CUE TITLE SLIDE, NAME, AND POSITION.)	Tannenbaum (SYNC): Especially in phramaceuticals, ours is one of the most policed industries in the world. There's no room for error.

In this example, the viewer is cued by the line, "What does exhaustive testing . . . mean to your customer?" Although the answer to this question should be forthcoming, the viewer is not certain of the source of the response. So, the one-sentence audio lead does not confuse the viewer as much as it creates a moment's drama. Who is talking? What are they saying? The customer (Tannenbaum) begins his response over an uncomplicated shot of a test tank in motion, thus permitting the viewer to concentrate on voice quality and what is being said. Momentarily, the appropriate visuals appear in sync and the image of Tannenbaum, a typical customer, is complete.

OFF-SCREEN AUDIO MONTAGE TECHNIQUE

Purpose

To use a succession of off-screen audio inserts to (1) predispose the viewer to accept a particular set of ideas or emotions (2) rehearse important elements presented earlier in a program and/or (3) give specific meaning and impact to corresponding visuals. The technique relies heavily on the capacity of the human ear to process and lend meaning to abstract information.

Description

A succession of off-screen sounds, scripted and recorded as the dominant element in an audio-video segment, can communicate with uncommon force. There are three main reasons for this. First, the source of the effect is off-screen. The viewer can imagine what or who the source is but is unable to certify that his/her hunch is correct. As long as the source of the audio remains off-screen, a hint of mystery and corresponding drama accompany the effect. "Whose voice is that?" "Is that a flight control tower or police dispatcher?" "Could you identify that noise in the background?" Clever blending of visual material with off-screen audio further enhances the emotional impact of a sequence, but even an awkward attempt at this technique attracts attention and triggers the imagination.

A second element that makes the Off-Screen Audio Montage Technique so useful is the multiplicity of audio sources. This series of audio inserts, one after the other, has a rhythm that involves the audience and holds it for the duration of the technique. Pacing *within* and *between* separate audio units can work the viewer towards the edge of his/her physiological seat or ease the audience slowly into a fade, dissolve or other soft transition. And because there is a series of different audio inserts, the technique can be used to imply credibility and wide acceptance of an idea or position.

Finally, there is almost never a time when the audience *expects* an Off-Screen Audio Montage. Without warning, the viewer finds himself/herself tracking screen action and integrating a lively-paced sequence of audio inserts, each one designed to contribute to a primary message. The "element of surprise" is part of the technique's structural make-up, regardless of specific program content. Very powerful at the head of a program for capturing attention and establishing context, the Off-Screen Audio Montage Technique is also an excellent interlude device, taking the viewer inside a character's head after an event to reveal how he/she thinks and feels.

A key advantage of this technique is its capacity to repeat a message without boring the viewer. Creative repetition of an idea is a hallmark of good design but is often difficult to carry out with certain media or media formats. The off-screen audio montage is so flexible technically that alternate forms of the same information reach the viewer as separate and different parts of a larger idea. Each "repeat" should be sufficiently different from the others so that the viewer processes it as a separate message. The mix of audio and visual combinations is extremely rich during an audio montage, allowing the production team near absolute control over what the audience will see and when they will see it. The traditional contextual constraints that come with a story line can be partially or totally relaxed during a segment when a succession of off-screen sounds dominate the message.

During production, audio inserts might be recorded at different times and in different locations, sometimes over the telephone. Drawn from myriad sources, these inserts are characterized by dissimilarities in background sound and overall ambiance. There are several ways to compensate for these sometimes annoying variations and avoid producing a montage that sounds choppy.

One option is to use a second audio track to lay down a consistent background sound for all inserts, for instance, room tone (hum) or music. Another is to cover the gap between inserts with a single sound that signals both an audio and visual transition. A common example of this is the sound of a motorized camera, used frequently to bridge from one still image to another. A third option is to electronically filter out most original background noise during inserts and cross-face audio tracks to eliminate noise between inserts.

In the editing suite, off-screen audio montages are best-handled by manually cutting and splicing 1/4" audio tape. The precision gained with manual editing vs. electronic editing is essential when the timing between two or more audio inserts and related visuals leaves no room for error.

With electronic editing, there is often a certain amount of "play" as the tape is moved across the audio heads, making it more difficult and time-consuming to locate the required edit point. For instance, the phrase "catch something" is difficult to edit electronically if the intent is to split the two words. As the words are spoken, the "s" sound in "something" blends with the "ch" sound in "catch." Electronically, it is difficult to position the tape at the precise point where the two words meet.

Using a manual procedure, the tape is moved by hand ever so gradually until the editor hears the break in sound between the two words. Scissors or a razor blade are used to make a clean, 45° cut and the edited segment is spliced against other program segments.

If electronic editing is preferred or necessary, tight edits are made easier by re-recording the tape at high speed, say 30" per second rather than the 7 1/2" or 15" standards. During normal speed playback, the length of the audio intervals is exaggerated, increasing the tolerance for error and making it less difficult to locate exact edit points.

Through the use of an Off-Screen Audio Montage Technique, non-specific visual material takes on definite meaning, and apparently clear visuals can take on an entirely new significance. For example, an auto worker, staring thoughtfully at his supervisor, hears a sequence of voices from earlier scenes encouraging him to file a formal grievance that would document the supervisor's unfair disciplinary practices. The voices are delivered off-screen at increasing volume and rhythm and begin to repeat as the tempo grows. The static visual, rather bland and nondescript in its own right, becomes a dramatic reflection of personal stress and uncertainty as the audio montage reveals the worker's inner-feelings.

Examples

The American Cancer Society employed the Off-Screen Audio Montage Technique in an award-winning public service announcement. The spot seeks to reverse the fear of cancer and create an optimistic attitude towards recovery. The central character in the 30-second announcement is a 65-year-old cancer victim.

The spot is divided into two segments. The character is pictured first in a series of dark-filtered shots, appearing depressed and listless. Framed by stark and barren backgrounds, the cancer victim appears ready to die. The audio portion of the program reflects his mood and symbolizes an attitude typical of cancer victims.

American Cancer Society PSA—Segment 1

Off-Screen Audio Montage: (Voice 1) It's no use. You've got cancer."

(Voice 2) "Give up. Give up."

(Voice 3) "No . . . ooo . . . ooo."

(REPEAT, REVERBERATE AND RAISE VOLUME)

The second part of the spot symbolizes the cancer victim's prospects for recovery. Here the character has undergone surgery. He tells us that he received a colostomy and is enjoying life to its fullest. He is pictured walking in a field of beautiful flowers on a sunny day. At the end, he breaks into a run and leaps into the air. This final shot is frozen, capturing the character in midair with arms raised and head skyward. The main character's voice-over is over-lapped by a second off-screen audio montage.

American Cancer Society PSA—Segment 2

Off-Screen Audio Montage: (Voice 1) "It's time to live again . . ."

(Voice 2) "You did it! . . ."

(Voice 3) "It's a new day . . ."

(Voice 4) "Nothing can stop you . . ."

(REPEAT AND REVERBERATE)

Imagine a second example. Early in a program on supervisory skills, the audience meets Helen, a model secretary. Her supervisors are seen collaborating and are heard discussing how hard Helen works and how lucky they are to have her working in their area. The visual message is consistent with their commentary; Helen is intent on her work and appears professional and efficient. A cut to Helen at closer range is accompanied by an off-screen audio montage consisting of three voices (Helen, Bob the Manager, and Mary the supervisor) and the intermittent sound of a typewriter carriage return. The off-screen message is carried by a series of conversational excerpts in which Helen requests new work opportunities, only to be put off by Bob or Mary. A sample excerpt is included below.

Video	**Audio**
CS on Helen working a WP station	Helen: Mary, I've been working here now for nine months without a chance to work on the new System 150. When I took this job, you told me . . . .
	Mary: Helen, Helen, you're right. It's just been a case of too much to do in too little time. Now, let me make a note and . . . .
	(SOUND EFFECT: CARRIAGE RETURN).

Three or four of these short excerpts establish Helen's disappointment over the insensitivity shown by Bob and Mary. The visual message is one of worker satisfaction and productivity. The off-screen audio reveals quite a different story with Helen, the prized worker, frustrated and dissatisfied by the lack of challenge and opportunity she faces. The off-screen message provides the viewer with significant information that Bob and Mary do not have. The viewer comes to see Helen as someone who maintains outward appearances, even as she contemplates options to her present situation. The program goes on to treat the issue of employee productivity and concentrates on "danger signs" that supervisors need to recognize and address.

USING MUSIC TECHNIQUE

Purpose

To establish mood, provide cues about relationships and interpretation of scenes, and to tie together program segments.

Description

Music is commonly used as an auditory accompaniment to the opening title or montage for training programs. Often, it is thought of as being primarily of aesthetic value rather than having specific instructional functions, although its applications are broader than typically represented.

When the goal of training is primarily affective, music can be a powerful tool. For instance, upbeat music with synthesizer stings or passages connotes progress and activity. It sets the pace for editing the visuals and can assist in actually changing the viewer's mood. Similarly somber music can heighten the impact of serious subjects, such as safety instruction.

Music can smooth transitions among edits and between scenes. For instance, when one or a series of interviews are edited together, there is generally a difference in audio quality, even among the various "takes" with an individual. Music in the background can cover over those differences, and provide a bridge between events. The ability of music to set a tone and to provide continuity is a special boon when using nonprofessional in-house talent. Often, it is necessary to do many re-takes, which causes changes in audio level and voice quality; sometimes these re-takes also leave the narrator sounding less than enthusiastic. It is amazing to note how that same person's tone is magically transformed when combined with an audio bed.

Music and sound effects can also be used as a non-verbal cue to attend to something specific on the screen or to engage in some activity. A beep or "oh-oh" sound effect can clue the viewer that something on the screen has been done incorrectly, eliminating the need for a wordy narrator. Similarly, music can signal a "flash of inspiration" or a sudden positive turn of events. In programs incorporating some sort of activity on the part of the viewer, such as answering a question in a workbook, or manipulating an actual object, a musical "signature" can cue them to do so, and another can cue them to return their attention to the screen. Sound can serve as an advance organizer: for instance, music can set the stage for "impending doom," so that even if the scene looks perfectly normal, the viewer is ready to expect a shift in tone.

Although not often used in non-broadcast television, custom "jingles" can be used to effectively impart a short message. This is especially successful in new product introductions which can use the techniques of television and radio commercials to have vocalists present the message and product name.

Examples

In a videotape on bank robberies for Marine Midland Bank, music is used to give the viewer a clue about the tone of the message as well as the characters who will be presented. The program opens with parallel scenes of a young woman and a man getting ready for work. The scenes use a synthesizer rendition of a classical piece containing two themes, one major and one minor. The theme in the major key is played when the woman is shown; the minor key is associated with the man. The action unfolds to reveal that the woman is a teller and the man is a robber who will enter the bank later. As the scene ends, we see the robber get out of his car, put the gun under his sweater, and approach the bank. On the last note of the music, a sound effect is superimposed which starts out as a high-pitched squeal decending to a vibrating crashing sound as the visual is held in still-frame. The effect of the music is to emphasize how an otherwise normal and pleasant day suddenly can be transformed by a robbery attempt, and to set the stage for subsequent very serious material.

In a second application, a utility uses a videotape to train customer service representatives in dealing with difficult customers. After each dramatized scenario, a message on the screen asks "How would *you* deal with this?", and a one-minute audio theme gives viewers both a cue and the time to think about a response.

OMNISCIENT SPOKESPERSON TECHNIQUE

Purpose

To use an "extra man (or woman) on the set" as a dramatic device to occasionally break continuity and provide viewers with analysis and explanation of important program segments.

Description

Although never acknowledged by talent, the omniscient spokesperson is an effective and inexpensive technique for any number of applications where program content can be broken into "chunks" for analysis, explanation or clarification. The all-knowing spokesperson shares ideas and insights with the audience, attempting to map-out a way of thinking about selected scenes. Behavior modeling programs are well-suited for this technique since the ramifications of positive or negative behavior can be examined immediately after (or before) they occur.

The Omniscient Spokesperson Technique is primarily a scriptwriter's tool. Since the spokesperson must be trustworthy, credible and knowledgeable, he or she is often introduced early in the program to establish rapport and authority. Setting up subsequent scenes and providing a frame of reference for the viewer help to establish the spokesperson's role early in the program. Once the first program segment begins, the omniscient spokesperson acts as a confidant, someone whose critique and advice are shared from an *insider's perspective*. The viewer should begin to feel that both he and the spokesperson are *co-evaluators*. Viewer reactions to a scene may be altered or reinforced by the spokesperson's comments.

Among the spokesperson's most useful tools are questions posed to the viewer either in advance of or after a scene. Used in advance of a program segment, questions prompt the viewer to focus attention on specific issues and to draw upon related experience and background knowledge. Used during or after a scene, spokesperson questions encourage examination of model behavior and arouse interest among viewers in their own behavior, i.e., performance.

Also useful are allusions to experience or opinions held in common by members of the viewing audience. They alert the audience to the spokesperson's familiarity with their work conditions, aspirations, concerns, etc. Here's an example:

> OMNISCIENT SPOKESPERSON: "Sure, Jack is having difficulty separating his personal problems from his on-the-job behavior. But we've all had some of that in our own work experience, now haven't we? Your personal life doesn't end when you arrive at work each morning. Mine never did. At least we know the problem is a common one. The question is, 'What can Jack do about it?' "

In this case, an appeal to the audience on a personal level is intended to draw them into an identification with Jack's behavior problem. If the strategy works, the audience will remain involved in the spokesperson's continuing analysis and subsequent identification of solutions.

When the omniscient spokesperson appears during a scene, talent can pantomime action or hold their positions. Sometimes, lighting is dimmed for dramatic effect and to focus attention on the spokesperson. As long as the spokesperson appears for a relatively brief period of time, actors in pantomime serve as a valuable prop. Actors can enter and exit pantomime routines on cue, permitting smooth and efficient segues to and from the spokesperson. References to and assessments of certain model behaviors can be exaggerated (emphasized) through the spokesperson's use of arm extentions, head movements and facial expressions. If the spokesperson remains on-screen for an extended period, talent can contrive movement, working around the spokesperson, but never acknowledging his presence.

Remembering that the omniscient spokesperson's credibility and therefore effectiveness depends on how well he/she relates to the viewer, every precaution should be taken to cast someone who has a look and presence compatible with the expectations of the intended audience. Unlike a commercial endeavor, where attractiveness, a nice smile and a sense of humor go a long way towards achieving audience respect, the task-oriented adult worker is often turned off by a look that is too polished. Keep this in mind during casting. Certain stereotypical qualities need to be considered as do particular characteristics of "leader types" in a specific business or industry. Generally, a spokesperson should appear older and more experienced than most of the viewing population. Delivery should be clear and precise, but industry jargon and references to past experience need to be blended into the dialogue. Here again, thoughtful scripting is the key to building an *image* for the omniscient spokesperson.

Example

In "Murphy's Law," a Factory Mutual Insurance program on Emergency Planning, "Murphy" is the all-knowing, all-seeing spokesman. Murphy pities mere mortals for their flaws but nevertheless guides the audience gleefully through a series of emergencies where a lack of adequate planning leads to predictable disaster. The point of each scenario is to relate the effects of the emergency to a collection of specific planning precautions which were unheeded and unattended. Had the planning steps been addressed systematically, the various emergencies would not have occurred.

As the omniscient spokesperson, Murphy recites his well-commercialized law, "If anything can go wrong, it will," and joyfully laughs his way through the emergency dilemmas. As he is transformed into a small, rotating star-shaped graphic to enter and exit scenes, Murphy actually plays an active part in contributing to the emergency situations (e.g., starting fires and tumbling containers). It's not that Murphy wants to create undue hardship for the characters he observes; he is simply duty bound to "teach them a lesson" the hard way by making sure they understand how inadequate emergency planning can lead to injury and loss.

The characters in "Murphy's Law" are never introduced as personalities, leaving Murphy to review their performance with the viewer from a perspective somewhat aloof and removed. The aim is to place emphasis on the importance of good emergency planning by observing what can happen in its absence and, at the same time, to ridicule those who fail to take the proper precautions. All the while, there is a tacit understanding that while the subjects of Murphy's critiques may be unaware or unconcerned about proper emergency preparation, they are probably no different than members of the viewing audience who are sharing vicariously in Murphy's "superior being" assessment.

As can be seen in the accompanying set of still frames taken from "Murphy's Law," courtesy of Penfield Productions, Murphy is a rotund, congenial chap whose British accent makes him all the more formalized as an authority figure. Note also that Murphy appears to be in his 50's, a desirable age for someone who has experience enough to communicate omniscience. In addition to the spinning graphic, also pictured below, the Factory Mutual program begins with Murphy on his own star, peering down on earth through space. This device used repeatedly between emergency planning segments, reminds the audience that Murphy occupies a place reserved for those with "powers and abilities far beyond those of mortal men."

Figure 3. Murphy, the fiftyish British spokesman.

Figure 4. Murphy perched on his own star.

Factory Mutual used a repertoire of production techniques to enhance the Omniscient Spokesperson Technique. The use of graphics and camera technique to position Murphy convincingly on top of his own star (see Figure 4), the spinning graphic that transports Murphy to and from scenes (see Figure 5), and galactic music that announces Murphy as extraterrestrial—these are three examples of other design and production techniques that Factory Mutual used to make Murphy a more convincing and effective omniscient spokesman. The point is that with rare exception, a dominant "theme technique" can be facilitated by other, less dominant techniques and video tools.

Figure 5. Murphy entering a scene with mere mortals as a spinning star from the heavens.

Still frame photographs courtesy of Penfield Productions, Ltd. in Agawam, Massachusetts.

OUT-OF-CHARACTER COMMENTARY TECHNIQUE

Purpose

To enhance the viewer's grasp of key ideas or events through periodic rehearsal, analysis and explanation of selected program segments.

Description

Out-of-character commentary occurs when talent step "out-of-character" to offer an analysis of their "in-character" behavior. Each commentary is directed towards the viewer and should focus on two or three important aspects of the surrounding program.

Out-of-character commentaries are supplements to selected program segments. Scripted to be brief and to the point, they should not rival actual scenes in either length or frequency.

For the viewer, out-of-character commentary has inherent appeal because it is novel. Actors who presumably deliver lines under the supervision of a director are transformed into expert analysts when they drop their scripted identities and address the viewer as "non-performers." There is a logical appeal as well: Individuals who have contributed to a scene as actors seem particularly well-suited to step back and discuss its implications as involved observers. Of course, both the program segment and out-of-character commentary are carefully scripted, but when talent deliver them in succession, the latter appears less staged and more personal.

On the set, the transition to out-of-character commentary needs to be acknowledged both visually and verbally. The general intent is to create a context for "honest" exchange, where talent momentarily leave the "dramatized" program to "talk things over" with the viewer. Close shots lend themselves to establishing the kind of visual intimacy that is necessary when characters talk directly to the audience. These can be completed as part of the program or through a slow zoom or several progressive cuts during the transition to out-of-character format. At the same time, talent must undergo changes in posture, facial expression and language to alert the viewer to a total shift in attitude from actor to observer. A shift in tone is particularly vital as it encourages the viewer to see the out-of-character personality as real and credible.

Although both the omniscient spokesperson and the out-of-character commentary utilize talent as on-screen audience representatives, the out-of-character commentary is delivered by an actor whose *primary function is to play a role in a program.* When he leaves his character to review the preceding scene, the actor's references to his own performance are introspective, while other scripted characters are seen from a detached, third-party point-of-view. His analysis is littered with remarks like "my first reaction was normal, but too hasty. I should have. . ." "What she was really telling me was. . ." and "Did you hear his response to my objection?"

There is no reason why two or sometimes three actors can't step-out-of-character to collectively assess their performance in a scene. More than three characters becomes unwieldy and too involved to handle efficiently. A two-person out-of-character commentary works quite well, since it permits interaction between different sources of information.

Example

Consider this adapted excerpt based on an actual program for licensed psychiatrists. It introduces a new anti-depressant and is produced and distributed by a major pharmaceutical company.

VIDEO	AUDIO
2-shot, office set	Dr. Paterson: . . . and so Carolyn, you're saying that, for the most part, depression occurs early in the day, usually just after you wake up?
CU of Carolyn, same camera perspective	Carolyn: That's right. If I can make it through the morning, things seem to improve. (Stress in voice.) At least, if I can concentrate on something other than how rotten I feel.
CU of Paterson, reaction shot nodding, then responding.	Dr. Paterson: I see . . . well . . . (Cut off in mid-sentence).

VIDEO	AUDIO
CU of Carolyn, slightly off camera from Patterson's perspective.	Carolyn: (Forcefully) Do you hear what I'm saying? I mean, is this kind of reaction normal for someone taking DoraFlex? Is there anything wrong with me? Dr. Paterson, talk to me . . . . do you understand or don't you?
MS, 2-shot	Dr. Paterson: Do I understand? (Turns to audience, smiles gently) That's a good question. Not just for me, but for any psychiatrist who prescribes DoraFlex to a patient.
MS of Carolyn	Carolyn: (Directly to camera, shift in tone): Absolutely! Remember, DoraFlex is a new drug. FDA approval came after 11 years of controversial blind and double blind testing. Some patients rave about its effects . . .
CU of Carolyn	. . . . . . but more than 60% have reported dramatic fluctuations in their reaction to DoraFlex during the first six months of treatment. So, when I ask Dr. Paterson (Hand gesture towards Dr. Paterson) if he understands, I really mean it.
MS, 2-shot	(Turns towards Dr. Paterson). Is DoraFlex right for me? Dr. Paterson: And I intend to answer you, but first I need to obtain answers to an entire inventory of diagnostic questions. Carolyn (gestures) has been taking DoraFlex for 11 weeks and she's reporting early morning depression. . . . not an expected reaction to this drug.

VIDEO	AUDIO
	. . .Do I need to alter the dosage? Should I prescribe a second anti-depressant for intermittent use? I can't make these kinds of judgments. . . . not until I complete the next phase of our session. Correct? (Turns toward Carolyn).
MCU of Carolyn	Carolyn: (Head turn from Paterson toward camera, nodding). True. And the strategy he uses to gather additional information is critical, both in terms
MS, 2-shot	of the quality and quantity of my answers . . . (Shifts postural direction towards Dr. Paterson) . . . Shall we proceed?
	Dr. Paterson: (Turns towards Carolyn, alters expression, adjusts clothing). Now . . . Carolyn . . .

SUBJECTIVE VIEWPOINT TECHNIQUE

Purpose

To establish a "camera's eye" point of view where verbal and visual cues work together to condition the audience to consider program content from a more personalized perspective.

Description

Unlike the more common third person camera position, where the audience observes the action from a detached perspective, the subjective viewpoint prompts the viewer to adopt the camera's point of view. The object is to establish a direct line of communication, albeit one-way with linear programming, between talent and viewer. When care is taken to create this "one-to-oneness," the program takes on a conversant quality that affects how messages are interpreted. By virtue of the vantage point adopted by the viewer, he or she is oriented to being personally responsive to program content. Worth noting is the fact that the subjective point of view can be used to condition the viewer to be someone other than himself, for example, a typical customer or a supervisor.

Describing the subjective viewpoint is far less complicated than achieving it. Camera placement, lens selection, camera movement, lighting, casting, scriptwriting and script interpretation must be skillfully integrated. The finished program must establish and maintain contact with the individual viewer on a personalized basis. The viewer needs to internalize program material in a way that goes beyond simple acknowledgment that "I'm part of this program." Very quickly, the viewer must begin actively processing information in a subjective and interpersonal fashion.

As "part of the program," the viewer is repeatedly encouraged to think and respond. "What do you think?" "I guess you've had that experience too." "Think about that for a moment." These are the kinds of verbal prompts that remind viewers of their involvement. Often shot in tight close-ups, talent deliver these kinds of lines directly into the camera. Sub-text (the *interpretation* a line should receive when delivered) is usually relaxed and conversational. Natural pauses and short intervals for reflection are important scripting points, providing the audience with sufficient time to think about what they've just seen and heard.

Talent's eye, facial and postural movement should be varied to maintain a conversational look. These language and body movement cues are vital in creating *intimacy* between talent and viewer and guarding against lapses into a talking head format. A knowing smile, a shrug of the shoulders and a friendly wink are the kinds of movements that ease the viewer into adopting a frame of mind that says "I like and understand this character. He seems honest and I think he knows what he's talking about."

With respect to lens selection, focal lengths between 10 to 15 millimeters most closely resemble the focusing characteristics of the human eye. With this degree of wide angle, good depth of field can be achieved so background objects remain clearly visible, even though a bit out of focus. When using this wide camera angle with a medium shot to medium close-up, the camera must be physically so close to talent (approximately four feet) that eye-travel can become a problem, even with through-the-lens-teleprompters. Professional talent who become familiar with the script should have little difficulty.

Inasmuch as the Subjective Viewpoint Technique is intended to lull the viewer into the camera's perspective, camera movements should be limited to those that the human eye is capable of executing. Cuts are permissible, since eyes can focus on discrete bits of information. Slow pans can be used to simulate eye movement across a flat plane. Zooms and other special camera effects, while commonplace in most programs, can shock the conditioned viewer out of the subjective viewpoint and are not recommended. Remember, as the viewer is being prompted by talent, there is a single angle from which all screen action is observed. Maintaining the integrity of the subjective viewpoint demands consistency in what the viewer is permitted to see and how he sees it.

The technique also calls for natural camera movements which change the location of the viewer, i.e., from a standing to a sitting position. Cuts are unnatural in that they omit the physical act of sitting down. Using a dolly in or out and a pedestal down most naturally simulates the act of sitting. A small studio crane provides uncanny realism in achieving the kind of effect that "seats the viewer." Likewise, any turn or tilt of the head, any kind of movement from one location to another, needs to be tested and blocked out in advance to determine how best to accomplish the move. In all instances, the intent must be to achieve a look that "feels" natural from the audience point of view.

Inserts or cutaways are permitted once the subjective viewpoint has been established. They are somewhat risky, since overuse of any type of camera mobility can produce an effect that is choppy. Verbal segues are necessary to condition the viewer for changes in screen direction or subject matter.

Finally, a producer must have a reason for opting to use the subjective viewpoint. Remember, the subjective viewpoint is not the same as a subjective camera angle, i.e., over-the-shoulder shot. Its effect is very different from what takes place when a spokesperson "talks to" the viewer by maintaining eye contact and employing a conversational tone. Rather, it is a *total conditioning process* where the viewer is an object of reference—part of the action. The attempt to manage this interaction between viewer and on-screen personalities hinges on anticipating how the viewer will respond to certain cues and how the program should respond in turn.

Certainly, the viewer can be scripted to be himself or herself. Under these conditions, a program is designed to simulate the real world and have the viewer respond accordingly. The question, "What would you do?" best characterizes this approach. For example, an audience of operations personnel in a nuclear facility can view a program from the familiar perspective of their normal work environment. An actor, playing the role of a co-worker, can condition the viewer verbally up to the point where an emergency occurs. Then, in an effort to trigger a mental response, the actor looks into the camera and asks "What would you do?" Anticipating either a correct or incorrect response from the viewer, the program continues with a depiction of ramifications, followed by a discussion of emergency procedures and recommended responses routines.

Often, a producer is interested in having the viewer assume the perspective of someone else. For example, a retail salesperson viewing a program may reflect on deficiencies in his own selling behavior if he is "cast" in the role of a typical customer. A staff manager who needs to improve her performance appraisal skills may see some of her short-comings in a model staff manager if she is conditioned to assume the role of a subordinate. Internalizing what they see on the screen, viewers reflect on their own behavior. "Is that what I sound like?" and "I hope I don't come across like that" are the kinds of thoughts that can go through a viewer's mind when he or she is cast in a participative role. Personal reflections can be discussed in detail when the subjective viewpoint is used in combination with any one of a number of techniques that explain or clarify. Several of these (e.g., Omniscient Spokesperson and Out-Of-Character Commentary) are described elsewhere.

Example

Monroe Systems for Business used the Subjective Viewpoint Technique in a sales training program entitled "Selling Simon: The 7869 Magnetic Ledger System." The viewer is informed that he/she will observe a model product demonstration "from the other side of the desk, here, where the decision-maker sits." A spokesman facilitates this conditioning as the model sales representative remains visible in the background. When the demonstration begins, the spokesman takes a seat adjacent to the one assumed by the viewer. A studio crane is used to lower the viewer into the chair and the demo begins. Throughout the demo, the camera pans slowly to the right to acknowledge the presence of the spokesman, who periodically comments on the sales rep's selling strategies. Always in contact with the customer/viewer but *never* acknowledging the spokesman's presence, the model sales rep maintains his concentration on presenting the ideal demo.

The effect of the technique is to transform an otherwise predictable, step-by-step review of features and benefits into a dynamic exchange of objections, questions and responses between sales rep Jack Williams and the spokesman. Actually, the spokesman is more of a "man in the middle" who at once represents the viewer through camera perspective and the customer through his off-camera questions and commentary. The still frames in Figures 5, 6 and 7 exemplify the positioning of both Williams and the spokesman as the transition is made from third party to subjective viewpoint. The accompanying script leads the camera towards its new perspective and is Williams' cue to turn and address both the viewer and the hypothetical customer.

Narrator: "What do I need to start using it? What equipment? What materials? What kind of forms do I need to use? Let's see how Jack Williams handles this one. OK, Williams . . . tell the customer what he wants to know."

Figure 6. With salesman Jack Williams in the background, the spokesperson begins to condition the audience verbally to assume the subjective viewpoint.

Figure 6, 7, and 8. Still frames courtesy of Monroe Systems for Business in Morris Plains, New Jersey.

Figure 7. The spokesman conditions Williams to address the customer (i.e., audience), as the viewer is slowly seated across from Williams and next to spokesman.

Figure 8. Once the audience is positioned in the subjective viewpoint, salesman Williams turns and starts his pitch.

CHARACTER TRACKING TECHNIQUE

Purpose

This technique is used to introduce two or more characters at the beginning of a program and then track their development through a series of interrelating events that influence characters, their behavior and its consequences. The emphasis is on character traits and interaction instead of event-oriented plot development. The technique is far more committed to "getting inside the character's head" than typical behavior modeling. With the Character Tracking Technique, characters are not simply vehicles that help communicate a message *about* content—*they are the content.*

Description

With this technique, every production decision is based on an interest in taking the audience one step further in understanding how a character(s) thinks, feels and acts. The assumption is that as a character becomes more clearly defined and better understood, the viewer sees more of himself/herself in the character and is more closely aligned with the character's approach to problem-solving, task accomplishment and performance.

The scriptwriter who elects to use the Character Tracking Technique must be ever-vigilant to the requirement that characters remain the central force in the program. Product information, selling strategies, management techniques and other content elements should not capture audience attention as independent topics. Factual information and ideas are significant in the ways characters manipulate or respond to them, but not as separate issues. If you choose to use character tracking, be sure your focus is on characters and only secondarily on events or information. This is the right technique if the seldom seen details of character thought and behavior are themselves the subject of the program.

Character tracking takes the viewer inside a character's cognitive and emotional world, attempting to plant one or two key attributes in the viewer's mind. Effective characterizations are believable. They prompt the audience to empathize and understand a character's point of view. While video alone seldom leads to a permanent change in behavior, it can be an "influencer" when the viewer and characters have similar objectives and operate under similar conditions.

The best use of character tracking threads two or three main characters together in a micro-examination of their motivations, behavior and interactions. This is accomplished by using different devices (e.g., subconscious voice, doubletakes, flashbacks, sequential repetition, shifting camera perspectives) to produce a compressed portrayal of interrelationships and common objectives.

To improve the likelihood that character tracking will have an impact on viewer performance, follow these guidelines:

1. The scriptwriter and director must manage character development to reach a logical conclusion that is justified by what the audience is permitted to learn about the characters.

2. Audience makeup must be consistent with the physical and mental attributes of the program's characters.

3. The viewing audience must be as homogeneous as possible in terms of work responsibilities, personal objectives, experience, etc.

4. Characters should be introduced early in situations (contexts) that closely approximate those familiar to the viewer.

5. Shot selection should include extensive use of closeups and subconscious voice to exaggerate character perspective and thought process.

6. No more than one or two themes or character threads should be used to carry the story line. Remember, this is not a soap opera with an endless pattern of plot and subplot. This is non-broadcast video intended for a limited audience with common objectives. The extraordinary concentration on character development will create interest; the focus on a limited message will improve the odds for transfer to the work environment.

7. Keep program length to a minimum. Introduced in the proper context for viewers with a common purpose, a 10-20 minute program can introduce characters, build credibility and communicate a message. There is an optimum length and the further you go beyond its limit, the more likely audience attention will wane or be diverted from the program's limited message.

8. The program must come "full circle" in an abbreviated period of time. Characters must be introduced, inter-relationships

established, and conclusions reached. Characters who begin discussing a last ditch sales approach for an unhappy customer should be rejoined at the program's end in time to evaluate outcomes.

9. Be liberal in moving ahead in time and space. With well-defined characters who "remind" the audience of themselves, the production team can rely on the audience to understand work-related conventions and procedures. Every change of location or action need not be explained. Use valuable time to reinforce how characters deal with issues in their pursuit of an objective.

10. Use gimmicks to exaggerate the focus on characters. Still frame dossiers displaying vital information (e.g., name, age, position, length of service with company, goals, recent accomplishments) and a photograph can accompany the introduction of each key character; or, similarly, a freeze frame can be used to immobilize characters while a character generator displays background information. Any device is useful if it introduces key characters as personalities and improves audience familiarity with character motives.

Example

A large publishing house has produced and sells one of the best competitive selling programs available. The program relies on Character Tracking Technique to compare and contrast the selling tactics employed by three different sales reps, each vying for the same major account.

The customer in the program is a publisher who is in the market for a state-of-the-art word processing system. Currently operating with a mixture of PC's, dumb terminals, and electronic typewriters, it is time to make the investment in a local area network that can accept information from multiple stations and integrate it for subsequent printing. For the three sales reps, this sale is a major opportunity and each one decides on a different strategy to obtain the account.

Once the audience travels through the customer's work area and meets two key decision-makers, the character tracking begins. Each of the sales reps is *tracked*: traveling to the account, arriving, meeting the prospects and executing his strategy. The story line follows an engaging pattern of segues, moving in and out of the three different sales calls. Comparative sequences capture each sales rep in basic phases of the systems selling cycle. The decision-makers give no clear indication as to which of the three products they prefer, so the audience is left to evaluate the action and make its own choice.

The Character Tracking Technique example reaches its denouement at an airport, where the three individuals, portrayed independently up to this point, make contact in the same scene. Standing in a waiting area, two of the sales reps strike up a conversation and discover they competed for the same publishing account. The two are profiled in conversation behind a row of seats. Hidden behind a raised newspaper and seated in front of the two reps is the third salesman.

When he realizes that the nearby conversation is about the same publishing account he visited, the paper is lowered and the torso is brought to attention. He clearly wants to hear more.

Sales Rep 1: "I thought I had it won . . . months of preparation, a competitive price and the publisher seemed ready to buy."

(SHAKES HEAD, LOOKS PERPLEXED)

Sales Rep 2: "That's the same feeling I had! The publisher was impressed with my package. But we didn't get the sale either. I wonder who did?"

(PAUSE AS BOTH SHRUG SHOULDERS CONFUSED)

". . . Gee, you don't think the managing editor was the 'real' decision-maker, do you?"

(CAMERA TILTS DOWN AND ZOOMS SLOWLY TOWARDS SEATED SALES REP)

(SEATED REP SMILES KNOWINGLY, STANDS AND EXITS TO BOARD PLANE)

With this final scene, the Character Tracking Technique is complete. Three different sales reps, introduced and tracked independently, meet at a common point in time and space. One is successful because he has read customer buying signals and sold the right person. Two go home empty-handed. While each has the benefit of only his own experience, character tracking has woven their separate activities into a vivid and informative comparison of sales techniques and buyer behavior.

ALTER EGO TECHNIQUE

Purpose

This technique improves understanding and shapes behavior by reviewing, clarifying or interpreting events and information.

Description

A technique that relies heavily on a well-written script, the Alter Ego Technique takes the form of a character's sub-conscious or inner self. Normally superimposed near a character's head, the alter ego interrupts screen action at key intervals to dispense advice and guidance. The alter ego can appear either as the character's "better self" who is physically represented by the character, or as a "trusted mentor" whose experience and dependability make him/her an ever present source of insight and wisdom.

The decision to use the same talent or a different character in the alter ego role hinges primarily on an analysis of the program's intended audience. The opportunity to treat the subject matter introspectively is greater when the alter ego takes the form of the character himself or herself. Viewers identify with the notion of "self" and draw parallels between their own work and that of the on-screen talent, who should be scripted to be very much like members of the intended audience. The character should be similar to members of the audience on key work-related dimensions such as income level, responsibilities, experience, etc. In these situations, the alter ego can be used to supplant certain thinking skills through rehearsal and logical analysis of information.

When a different character is cast in the alter ego role, he or she is in a better position to portray the voice of experience. Obviously, this form of alter ego is not physically representative of the inner self, but information is clarified and guidance is provided by a source that resides in a character's sub-conscious. Audiences with lower experience levels and/or less on-the-job authority are better targets for this version of the Alter Ego Technique since its contact with the "self" is more didactic and directional. Unlike the "better self" who interacts with a character to analyze and reach decisions, the "trusted mentor" is largely a teacher and coach who is interested in suggesting how the character should think, react and behave.

In either case, the alter ego interrupts screen action to consult with the "self"; the visual context of the interrupted scene remains as background during their exchange. As a result, the alter ego does not engage in extended dialogue. Instead, he or she appears periodically to focus on a specific decision, task or problem. Events are summarized, information is clarified, options are considered and recommendations for action are made. This permits the viewer to observe action leading up to the appearance of the alter ego, to consider how he (the viewer) might react under similar circumstances, and to return to the interrupted scene where the "self" tests the viability of the alter ego's input. The viewer sees the character under performance conditions where the consequences of his/her behavior are readily apparent.

When the scene is interrupted, action may or may not be frozen. Characters other than the "self" may continue in pantomime while the alter ego is on-screen. For example, a manager who is faced with an irate employee may make contact with his/her alter ego even as the employee continues to issue complaints. As the manager is mentored by the alter ego, the employer may continue to pantomime irate behavior. These non-verbal cues maintain continuity throughout the scene and permit the alter ego to appear naturally as part of the "self's" thought pattern. In this case, the manger remains part of the pantomime, but at the same time, is visually in contact with the alter ego who is explaining how the employee's grievances should be handled.

The option does exist to freeze the action and pick it up once the alter ego has completed his/her lines. Lighting can be dimmed while the alter ego is on screen to eliminate potential sources of distraction. Freezing the action works best when the alter ego reminds the "self" of past experiences or other sources of information that need to be considered; these can be superimposed as part of the "self's" sub-conscious. The viewer is expected to focus attention on both the alter ego's remarks and the introduction of visual information from another time and place. Under these circumstances, it is best to freeze screen action momentarily and return to it with a transition that reacclimates the viewer to the ongoing scene.

The alter ego is a special effect that can be inserted live or during editing. It should be positioned near the "self" so that contact can appear direct and personal. Eye-to-eye contact between alter ego and "self" is an important element when the alter ego takes the form of "trusted mentor," since the character is physically different and inclined to look at the object of his advice. When the alter ego is used several times during a program, it can be positioned in a variety of locations to reduce monotony and complement frame composition.

When the effect is inserted live, a dedicated camera is focused on the alter ego character. A long shot is required to control the size of the alter ego's head and make it compatible with the relative size of the characters in the ongoing scene. If the alter ego effect is shot full-screen and compressed during post-production, shots can be picked up conveniently, although the physical movements (i.e., nods, raised eyebrows, smiles) of both the alter ego and "self" becomes more difficult to coordinate. In either case, the inserted effect is keyed in and is enclosed by either a soft frame or border. It can be introduced with a fast wipe, appearing visually to pop into the action, or with a slow, hazy dissolve that alerts the viewer to upcoming contact with the inner self or subconscious.

Example

In the Professional Development, Inc. (PDI) program series, "The Making of a Salesman," a white-haired sage appears periodically to discuss the selling process with younger sales representatives and managers. This senior advice-giver and mentor is named "Willie," a loose reference to the symbolic Willie Loman in *Death of a Salesman*. Like the Arthur Miller character, PDI's Willie has seen it all and done it all. Whatever the question or problem, whenever it occurs in the sales cycle, Willie has already made the mistakes and learned the valuable lessons. PDI's Willie, however, is neither troubled nor bent on self-destruction. He embodies all that can be good in a veteran salesperson and is available to show others the way to success.

Willie shares lessons from his experience with various characters throughout "The Making of a Salesman," drawing the floundering or unsure salesperson into brief assessments of events and performance. The pattern in each encounter is the same. Willie intervenes whenever his protege faces an important decision, usually on the heels of a disappointing customer interaction or just as an opportunity to enhance the sale or customer relationship presents itself. Each encounter includes three components:

(1) identification of the issue;
(2) analysis of performance, both good and bad, and a review of the effect this performance has or is likely to have on the customer and the sale; and
(3) direction and guidance for future action.

As the alter ego, Willis is reassuring, positive and supportive. He has no interest in personal gain except for the satisfaction he receives from helping the younger, less experienced sales rep to succeed. Intended for developing sales representatives, the PDI vignettes portray Willie as the mentor and advisor most salespeople need but seldom have access to.

The photograph and accompanying script in Figure 9 are taken from a scene where the sales rep (Danny) is attempting to develop a relationship with a difficult purchasing agent (Fowler). Danny has made a few mistakes in attempting to build a new relationship with Fowler, but he recovers by following Willie's advice and the systems selling strategy espoused in the program. In the scene captured here, Danny has just received the go-ahead from the purchasing agent to submit a proposal.

Willie: "You did OK Danny Boy. Instead of blowing your cool, you started in to pacify a guy you need on your side and it paid off. Now you've got a solid working relationship going that started on shaky ground. Remember, there's a lot of Fowler's out there, so know the buying procedures of each of your customers and develop good rapport with all key company buyers."

Figure 9. Still Frame from "The Making of a Salesman" used with permission from Professional Development, Incorporated, Cleveland, Ohio.

PARALLEL SCENES (ACTION) TECHNIQUE

Purpose

To relate key action from two different story lines, separated by time and/or space, in order to:

(1) show cause/effect relationships;
(2) compare/contrast ideas or behaviors;
(3) omit unnecessary details; or
(4) show simultaneous events or people who will eventually "meet" in an upcoming scene

Description

Parallel action is a highly interpretive technique that leads the audience through two (or more) story lines, intermittently jumping from one to the other as key relationships are established. Parallel action is not meant to show one scene following or preceding another; the intent is to connect two sequences for purposes of subject matter association. The technique takes advantage of what is perhaps video's greatest inherent strength—the capability to transcend time and space.

Each new story segment represents part of a progression of events. Audience attention is maintained, in part, through the periodic shifting from one story line to the other. But parallel action is much more than simply cross-editing two scenes. In order for the viewer to be emotionally involved in tracking program content, between-scene and within-scene pacing must be planned, controlled, and coordinated. Jumping back and forth rapidly between story lines can create a sense of excitement, but ends up confusing the viewer if the scenes themselves fail to maintain a similar level of vitality.

The script should be written to accommodate the side-by-side development of parallel scenes. Each separate segment should focus on a single message, thereby controlling length and helping the viewer make associations on a point-by-point basis. There needs to be a closely knit arrangement of audio and visual cuing in each story line to alert the viewer to what's coming and what has already taken place. These cues aid the audience in tracking and integrating information.

The key throughout, is to create the impression that parallel events are ongoing and, while it is not necessary to show each story line in its entirety, to ensure that salient information is provided. It's as if the audience is shifting back and forth between scenes so as not to miss any important action. There should be a feeling of involvement and increasing interest in tracking events as they move progressively towards a conclusion.

This does not mean that parallel editing must be fast-paced. Especially in programs where dialogue is central to the message, pacing must be slowed to ensure that the viewer understands the audio content. Through a combination of production factors (e.g., frequency of camera movement, number of edits and camera switches, length of speaking parts, pitch and duration of sound effects, etc.) each segment acquires its own tempo and contributes to the pace of the entire parallel sequence.

Parallel action offers two additional advantages. First it increases the amount of useful information that is transmitted in a given amount of time. Extraneous and irrelevant material can be omitted because the script is structured as a series of segments that are intended to portray selected information only. Second, it eliminates the need to offer peripheral explanation via a spokesperson or titles. The relationship between the two scenes is organized and develops gradually, improving the viewer's ability to see and interpret relationships.

The periodic movement between scenes is accomplished through various transitional devices. Each is different and can be preferable for a particular kind of parallel development. Once a transitional device is selected for use during a sequence, it should be used exclusively. Using several different devices serves no useful function and can easily confuse or irritate the viewer.

Of the many transitional devices available, seven are discussed below. These are: straight cut, rapid fade, rapid dissolve, split screen, wipe, digital squeeze, and whip pan.

Straight Cut. The simplest transition and one that is particularly well-suited to parallel action, the cut signals an immediate change from one story line to another. It induces the viewer to instantly associate two events or ideas. With a cut, the transition is sudden; the audience is pulled from one situation and thrust into another.

This crisp movement between story lines keeps the audience involved in tracking the action. Before shooting, it is important to review the number and frequency of all cuts. Too many cuts *within* scenes can compromise the effect of transitional cuts *between* scenes.

Rapid Fade. Less dynamic than a straight cut, the rapid fade is useful when parallel scenes are separated in time or time and space, but are still associated in content. Even a fast fade takes time and fading to and from a black screen creates a momentary pause in program flow. The viewer interprets this pause as a sign of temporal or spatial change, signaling that the two scenes are non-concurrent.

Rapid Dissolve. A gradual dissolve is normally used to indicate spatial or temporal change, especially the passage of time. Interestingly, a rapid or fast dissolve can suggest concurrent action. The intermittent and rapid blending of one image into another creates the impression that the action in both locations is simultaneous and continuous. The viewer's natural inclination is to follow the transition and see how these two situations relate to one another.

Split Screen. Unavoidably, the split screen is a vehicle for portraying parallel action. With the screen divided, two simultaneous events can be compared, contrasted or simply seen together. Often used to dramatize direct interaction between events (e.g., two parties engaged in telephone conversation or two athletes engaged in the same sporting event), the split screen has limitations. More complex relationships between separate events can be confusing when two sets of visual and aural cues compete for audience attention on the same screen. Also, a divided screen permits only a portion of each contributing scene to be used; restrictions in the size and proportion of shots make it unsuitable for many full-screen applications. As a rule, split screens work best with uncluttered visuals (sometimes static) and a single audio source, usually voice-over narration or background music.

Wipe. Using an analogue-based production switcher, a wipe appears as an electronically-generated line or shape that removes one scene and covers it with another. Using a digital switcher, one scene can be "pushed off" while another is "pulled on." With the more expensive digital wipe, images appear to glide in unison, one pushing the other out of view. Good scriptwriting takes into account the brief period of time during a wipe when both scenes appear at once. Action in the outgoing scene "gives way" to the replacement scene, permitting the viewer to transfer attention smoothly from one visual image to the other. By alternating the direction of the wipe, you can graphically signal the transition from one story line to another. Left-to-right wipes could signal transitions to story line A, while right-to-left wipes could signal transitions to story line B.

Digital Squeezing. Digital squeezing can compress pictures, transporting a full-screen image to a quadrant or other area of the screen, even reducing it to the size of an indistinguishable dot. As the original image is squeezed, a new one takes its place, communicating its message to the audience. In order to portray parallel action, the compressed image is brought again to full screen as its counterpart is squeezed. During the process, an image can be "flipped" or "spiralled" to lend dramatic and entertainment value to the transition. Digital squeezing is more expensive to produce than straight cuts, rapid fades or other analogue-based techniques and does not enhance the viewer's ability to track parallel action. When a target group expects or can be influenced by broadcast quality production, however, digital effects are worth the added expense.

Whip Pan. Also referred to as a "blur" pan, this effect appears as a series of rapidly moving, indistinguishable images moving either left-to-right or right-to-left. This streaking effect creates the impression that a temporal or spatial transition is taking place. The effect is appropriate for most cause-effect relationships, comparisons, shifts in perspective, etc. It captures attention because it is different, confusing and fast-moving. As with a wipe, panning in one direction can signal a change to story line A, while a pan in the opposite direction would represent a transition to story line B. The whip pan should be used sparingly, however, since overuse can focus attention on its blurring images instead of the program's subject matter.

Example

A security alarm company video uses parallel action to train tellers in the procedures they should follow during attempted holdups. The program begins by following the parallel activities of a model teller and a would-be robber as they prepare for and carry out a day's work. Once the teller and robber "converge" at the bank, parallel action gives way to a single scene where information about the alarm system is isolated and reviewed in a series of freeze frames. Here are storyboard visuals from the teller training video where a whip pan is used to segue from Parallel Series A to Parallel Series B.

Parallel Series A: Robber driving to bank

Robber leaves house, enters car and begins driving.

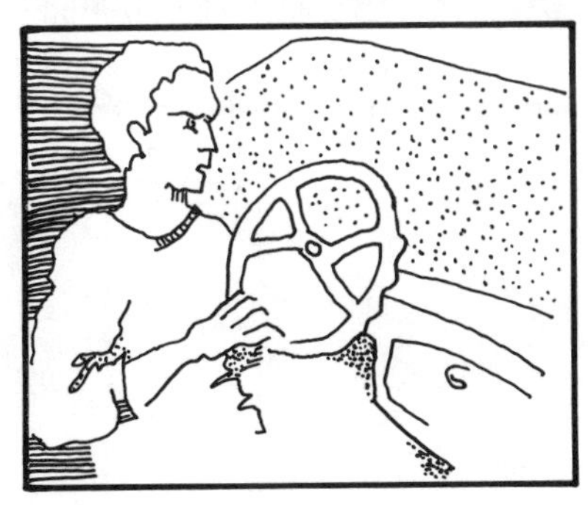

Robber in car, fidgeting, listening to radio.

Parallel Series B: Teller going to work.

Teller leaves house and walks to bus stop.

Teller sitting on bus, chit-chatting with other riders.

Robber at red light waiting nervously, impatiently . . . waiting for pedestrians to cross street.

Robber pulling car into store. Enters store to make purchase.

Robber returns to car, smiles at radio announcement about low crime rate.

Duration: 90 seconds

Teller looking out bus window, smiling in good spirits.

Bus stopping at corner. Teller exits and walks last block toward bank.

Teller enters bank.

Duration: 90 seconds

QUICK VISUAL INSERTS TECHNIQUE

Purpose

Often used in context to support verbal information, this series of quick visual cuts provides the viewer with concrete representations of objects and events.

Descriptions

Although quick inserts can be used at any point during a program, they are frequently part of introductions and summaries. In the case of the former, short visual segments are cut into the program to plant well-defined pictures in the viewer's mind. This is a particularly useful strategy when the *early part* of a program is intended to (1) recap information already presented as part of other programs in a series, (2) prompt viewers to recall certain elements from past experience which need to be accessed as prerequisites to the presentation of new information or (3) plant key images in the viewer's mind before they are treated in the context of a program.

When selected images are inserted near the *end* of a program, they help the viewer retain important information. Whether the images are identical or similar to others appearing in the body of the program, the fact that they are repeated increases the likelihood that the viewer will see them as essential and meaningful. In many cases, these selected images are the central elements in the program around which all other material is organized. Used in the *body* of a program, quick visual inserts produce strong images in support of the ongoing story line.

Whether quick visual inserts are used to establish a mind-set for subsequent information or to emphasize key concepts and ideas, their message should be clear and easily discernible. Each image should appear to "pop" on the screen. The viewer should have the feeling that the insert "jumped out" with a vivid and recognizable message. Different views of the same subject matter and particularly dramatic or imaginative shot selections can increase the likelihood that the viewer will retain the image. The sequencing, frequency, location and length of each insert should be determined in advance to heighten impact. In terms of program context and organization, each insert should "make sense" in terms of subject matter continuity and contribute to the program's flow.

Scripting must anticipate and be built around a series of quick cuts to account for and take advantage of their abruptness. Pauses, concise dialogue, and verbal reminders are effective devices in ensuring that the viewer concentrates on visual inserts and associates them with related verbal information. During production, the director must capture the essence of the narrative and shoot insert material that can communicate its message fully in a fraction of a second. Several scripting devices are used in the following excerpts to enhance the summary powers of the visual insert technique.

LS Patient group seated in semicircle, engaged in discussion.	VO Spokeswoman: . . . And isn't that what self-help is all about? People learning to help themselves?
CS Martha, smiling, then speaking.	Remember Martha. A fifty-year old alcoholic. Recovered today with help from someone like you.
CS/Reverse angle. Slow tilt up from Martha's hands in lap, rubbing back and forth across hankerchief, to CS of face.	Someone who was willing to donate three hours a month to help a fellow employee. (Pause) I can't tell you how much the Self Help Program has meant to Martha—only she can tell you that.
MS Spokeswoman (Limbo set)	SYNC Spokeswoman: (Begins slow, deliberate walk across set) But I can tell you that more than 250 Beecher employees have received Self Help services over the last three years.
CS Spokeswoman (Completes line, turns left)	And I can tell you what Self Help could mean to you or a member of your family.
CS Spokeswoman (Matching action)	Think about some of the people you've seen today. Think about Raymond.

MS/Visual Insert Raymond at work on line	VO: A line supervisor. Fifteen years with the Beecher Company.
MS Spokeswoman (Pan as she stands and walks left)	SYNC: Only 44 years old, Raymond loses his wife to cancer and suddenly, for the first time in his life, Raymond is alone.
LS Spokeswoman walking toward camera slowly	SYNC: Think about Jerry Samms.
CS/Visual Insert Jerry seated on living room floor, playing with toys	VO: Born with a respiratory ailment . . .
LS Spokeswoman continues slow walk towards camera	SYNC: Restricted to a life indoors where air quality must be controlled and monitored. Some of you know Jerry's mother . . . she's your co-worker. Some of you also know that Elaine Samms has been a Self Help volunteer since she came to Beecher three years ago.
CS/Visual Insert of Janet behind desk working. (Spokeswoman turns left to smile)	VO: And who can forget Janet?
Same view of spokeswoman walking towards camera (Tilt up to waist, then to head and shoulders)	SYNC: On disability for six months with a broken hip . . . and when she was able to return to work . . . her job had been eliminated. But Self Help supported Janet with a training program . . .
CS/Visual Insert of Janel being trained	VO: One that led to a better job . . .
Same view of spokeswoman (CS as she completes walk next to Beecher Self Help logo)	SYNC: and more responsibility

CS of spokeswoman (Looks off to logo, then to camera)	SYNC: Now what about you?

The script for this program continues with an appeal to all Beecher employees to donate time to the company's Self Help program. Of course, the viewer has already met Martha, Raymond, Jerry and Janet by the time this recap takes place. The details of their situations and the assistance they received have been reviewed. The quick cuts to these people who are recipients of Self Help services are a natural way to remind viewers that by volunteering time, they contribute to a fellow employee in a very real and tangible way. In this case, visual inserts serve to dramatize the human factor so essential to soliciting volunteer services. The verbal appeal for pledges comes on the heels of the inserts and precedes a live appeal immediately following the video program.

Examples

Savin Copiers uses the Quick Visual Insert technique in a promotional program that introduces its dealer organization to new additions to its copier line. A spokesman walks onto a dimly lit set. The only audible sound is that of his footsteps. He removes his coat and begins talking about the copier line. As he does this, the set is lit slowly, one light at a time. Interspersed in the dialogue are a series of quick inserts, each one displaying a Savin copier in use in a business location. By the time the spokesman has completed his overview of the existing line, he has taken a seat near the front of the now fully lit set, where he can begin his introduction of the new products.

The Columbia Bicycle Company demonstrated the versatility of the technique in a program about their assembly process. Figures 10, 11, 12, and 13 show how Columbia used quick visual inserts to transport the viewer from front office to shop floor.

Figure 10. The president of Columbia Bicycle leads with 90 seconds of straight narration. The audience becomes familiar with his appearance and delivery before the first in a series of quick visual inserts replaces him on-screen.

Figure 11. Woman applying pinstripes (visual insert).

Figure 12. Man transferring frames from conveyer to packaging (visual insert).

Figure 13. Woman rotating wheels (visual insert).

These three still frames and several others are actually used to segue from the president's on-camera narration to voice-over and the bicycle assembly process. The visual inserts appear more frequently and for longer durations until the transition to documentary is complete.

Figure 10 shows how the Columbia program opens using the company president-as-spokesman in a talking head format. As the static shot of the president continues, his verbal references are supported through a series of quick visual inserts. Figures 11, 12 and 13 are three in a series of visual inserts that grow in both frequency and duration as the spokesman continues his opening remarks. The last visual insert marks a transition point from on-camera to off-camera narration with the company president continuing to provide voice-over. The visual continuity is restored as live action shots from the shop floor begin to document the bicycle assembly process. The effect is almost subliminal as the audience is eased from talking head to documentary format. In this example, the Quick Visual Inserts Technique is used to (1) offset an otherwise uninspiring opening with brief shots of workers in action and (2) segue creatively from on-camera narration to voice-over and the bicycle assembly process.

FLASHFORWARD TECHNIQUE

Purpose

This technique provides the viewer with a premature glimpse of future events by altering the chronology of a story line. The intent is to draw an early relationship between what may or will take place later in the program depending on interim events and behaviors.

Description

Despite wide acceptance of the "flashback" technique, where the viewer is carried back in time to relive an event, there are not nearly as many examples of the technique being used in reverse, where the viewer is transported ahead in time to see the future.

In video programming, as in life itself, people learn from experience. They keep track of what has happened and what is happening in order to anticipate what will happen. A viewer's tendency to think "in chronological sequence" does not mean that he/she should always be treated to a story line where outcomes are uniformly revealed after the events that precede them. There are situations where a program's impact is enhanced by interrupting the chronological flow of events to present outcomes which may occur at some later point in time. Merely "knowing" what will happen before it takes place is itself an intriguing prospect for the viewer.

Although the flashforward can no doubt be used occasionally as a creative tool, it is generally appropriate as a functional strategy when:

(1) The rewards tied to performance are novel, excessive or arranged differently, making them a paramount issue in the program;

(2) The risks associated with unacceptable performance are so serious that potential outcomes and consequences require special attention; or

(3) Achieving a goal depends on motivating viewers to anticipate rewards and self-satisfaction as a by-product of their contribution and commitment.

These three avenues for concept and script development are based on a common theme: communicating expected outcomes clearly and with impact is vital to meeting program objectives. Even though events leading to outcomes are integral to program development, it is the outcomes themselves that are apt to cause a significant change in the way people think, feel, and perform. The creative team responsible for the program has the option to use the flashforward as (1) an absolute predictor of what will happen or (2) a behavior modeling tool for presenting several alternative views of what could happen.

As a strategy for shaping performance, flashforwards encourage the viewer to think about consequences before deciding on a course of action and then to watch vicariously as models exhibit either appropriate or inappropriate behaviors. What the viewer sees in a flashforward may or may not be an actual portrayal of final outcomes. While three possible outcomes might be portrayed early in a program, a character's behavior and performance can determine which set of consequences is warranted in the end. Even if the flashforward is used sparingly, it places the viewer in a position to say, "I've seen what the likely outcomes are. Now let me understand the events that lead to these outcomes." Ultimately, of course, the viewer should carry this thinking a step further and decide how to perform in order to produce the desired outcomes for himself/herself.

Whenever the Flashforward Technique is used, two production requirements should be met. First, the script should contain language cues to signal the transition into and out of the flashforward. The audience needs to be conditioned verbally to recognize that the program's chronology is being reordered. Second, the transition from the present to the future should be signalled visually. A defocus dissolve or ripple dissolve is normally used to denote an imagined change in time or space; either can be used for flashforwards, flashbacks or dream sequences. Both create a hazy, dizzying image, not associated with more frequently used transitional devices. They signal a departure from the present and are necessary to both enter and exit the imagined segment. (Note: A whip pan is sometimes used to signal a transition ahead in time. It's blurred series of hardly distinguishable images creates a sense of jumping ahead to a different scene. All of the visual devices used with flash-forwards can be accompanied by music or sound to enhance the feeling of movement.)

Flashforwards can occur at any point in a program from the first scene to the last and anywhere inbetween. It's important to consider how the positioning of the flashforward can influence the audience. A flashforward appears at the head of a program to gain attention and set the stage for dramatic development and continuity with closing scenes. The technique is used in the body of a program to reveal potential consequences ahead-of-schedule, making sure the viewer focuses on the ramifications of his/her performance. At program closing, a flashforward provides a vision of the future—an image of a goal waiting to be reached.

A fine line distinguishes the flashforward from a kind of advance organizer that is used to gain attention in informational programs and documentaries. A program on fire safety may begin with a series of clips, each graphically portraying the tragedies associated with fire. The clips organize the audience's thinking in advance for the remainder of the program, which is oriented to general information on fire prevention. But since the precautions reviewed in the body of the program are general, and not related directly to the consequences, i.e., tragedies portrayed up front, the flash-forward technique really isn't being used.

Example

The excerpt below includes both verbal and visual cues as a flashforward takes the viewer from present to imagined future.

Video	Audio
2-Shot/side view of Joe and Bob	Joe: Bob, I want to make sure you understand this new bonus program. It goes into effect January 1. (Hands copy of bonus plan to Bob)
MS of Bob	Bob: (Takes copy of plan, looks at it casually) Don't worry about it Joe. I read the memo from Corporate, most of it anyway.

MS of Joe	Joe: Oh yeah? Did you see the section on double bonus points?
CU of Bob	Bob: (Caught off guard, obviously interested) Now, hold on a minute. Double bonus points? Are you serious?
CU of Joe	Joe: You'd better believe it. Once you clear 130% of quota, this new plan kicks in . . .
Reaction shot/CU of Bob, Slow zoom-in	(Bob tilts head and smiles whimsically as he listens) . . . Imagine how much you could make if you replce 3 or 4 inhouse systems with new XP83's . . .
CU of Bob/Defocus Dissolve to MLS with Bob on podium receiving award	Joe: . . . I mean Bob, with your territory, this is an opportunity to hit six figures . . . think about it . . . not only the money but the prestige, the recognition . . . maybe even a promotion . . .
Flashforward/LS of Bob receiving award from CEO	CEO Hendricks: Bob, as this year's outstanding sales performer, I have the privilege of presenting you with this check for . . .

In a not-so-different example, a promotional program designed to raise funds for restoration of the Statue of Liberty begins with scenes of famous national landmarks in various stages of disrepair. The Statue of Liberty is the final landmark, portrayed completely restored as seen in profile from the air. Patriotic music and the sound of cheering and applause accompany the scene. The program goes on to review the efforts and costs associated with the restoration and then returns to the profile view of the celebration. Shot from a helicopter circling slowly around the statue, an opposite profile view soon appears. A pock-marked and damaged statue fills the screen; there are no signs of life and the only sound is that of a hollowing wind moving against empty scaffolding and support cables. The program ends with a close-up freeze-frame of the statue's face, still from the unrestored side, with a title bleed which says "Please Give."

This is an example of a flashforward. It offers a vision from the future where a celebration is underway to mark the finished restoration. It returns to provide information on how the statue is being restored. It concludes by returning to the imagined celebration as a reference point, but it also flashes forward again to portray the other possible outcome—a statue that is only partially restored.

GRAPHICS AND ANIMATION TECHNIQUES

Purpose

Graphics and animation are used to "distort" reality so that a concept can be more readily understood, or to provide images of things that don't exist or can't readily be captured by a camera and videotape.

Description

Often, the role of training is to explain some concept or phenomenon that can't readily be perceived. Graphics and animation can compensate for our inability to perceive things which are too large, too small, happen too quickly, or too slowly. Such artwork, produced by traditional means, or more commonly, by computer generation, can also allow the viewer to "see through" objects, or to more easily pick out the salient portions of an otherwise confusing scene. Common examples of these applications are graphics and animation used to explain scientific and technical topics, such as the flow of electricity through devices, or the rotation of gears and motors within machines.

Often, viewers can become overwhelmed with too much detail in a realistic depiction of an image; in fact, it is not the case that the more realistic an image, the more effective it is for learning. A novice may have difficulty in picking out the important details in a shot, and even if the camera zooms in on a tight shot of the portion being explained, the context may not be clear. A graphic can simplify the image, leaving out extraneous details such as knobs, buttons, and wires, that are not being discussed as yet. Simple line drawings using color and perhaps labels to highlight the topic being presented can aid in focusing attention. Graphics and animation can also distort reality by providing "exploded" views or magnifying areas. Using many of the techniques made easier by 3-D computer graphics generators, an object can be rotated, turned upside down, and made transparent as the viewer can seem to be able to travel through its exterior shell and explore the inside. This technique is a spin-off of CAD-CAM (computer assisted design—computer assisted manufacturing) software which allows such manipulations of images. It is commonly seen in commercials for cars and in architectural presentations.

Graphics and animation can also be used to create characters that rectify concepts or serve as actors. This technique can give the viewer a visual referent for a concept or process, and can facilitate the depiction of human behavior without linking it to a particular person, age, sex, race, etc. For instance, physical or electrical forces might be personified as teams or armies of individuals; another popular example of some decades back was Redi-Kilowatt, the animated character developed by General Electric. When it is desirable to portray undesirable behaviors or to exaggerate behaviors, it is often advisable to use animated characters rather than live actors who may look too unrealistic or too unattractive. Exaggeration and humor can often be accomplished more successfully by such obviously UNreal characters. Another advantage of animated characters rather than live actors is that they can be drawn to depict a "generic" being that is ageless, sexless, raceless, and never out of date. When one looks at programs produced even a decade earlier, they often become ineffective because the viewer's attention is drawn to extraneous details such as hairstyles and clothing styles which may be dated. However, even 20-year old animation looks "current," as Mickey Mouse has proven.

For the presentation of numeric or statistical information, charts and graphs are an old stand-by. If, however, they are overused, the presentation can look more like a slide-show, and in fact, might more appropriately be produced in that medium rather than video.

Graphics and animation can be costly and time-consuming to produce; computer-generated animation is commonly estimated (in 1989 dollars) at one thousand dollars per second, and even hand-drawn art-cards easily cost $250. However, there is rapid development in the area of computer graphics specifically for video production, and as the price drops and ease of use increases, it will be common to find these devices even within modest in-house video studios. An advantage to producing and storing computer graphics rather than traditional paper-and-pen images is that computer graphics can easily be manipulated and altered. For instance, colors, sizes, and text can readily be modified. If images are to be used again in a subsequent program, updated figures can be inserted almost as easily as making word-processing changes. Complicated logos, characters, or drawings can be stored on computer disk and re-used like "clip art" for a number of purposes.

Example

Reliance Electric uses computer animation to explain many mechanical and physical principles of the devices it manufactures in its introductory training of sales engineers. Examples of that use are contained in a program on the basics of gearing. To define the term "horsepower," the narration explains that one horsepower is the energy required to lift 33,000 pounds one foot in one minute, or to lift one pound 33,000 feet in one minute. A visual analogy is made by computer graphics showing a building being raised one foot and one brick from that building being raised 33,000 feet. Later, to explain the principle of a gear, a lever is shown rotating a wheel; that lever is then duplicated around the circumference of the wheel, becoming, in effect, teeth of the gear and multiplying the torque. These effects were produced using computer animation.

Monarch Tools manufactures computerized machining equipment. To explain the operation of the various moving parts that cut metal, the program dissolves between perspectives of the real machine and a hand-drawn simplified machine. Using clever cardboard cut-outs manipulated from the back of the graphic card, the artwork can be animated to show the movement of the device in slow motion.

These two examples are drawn from programs about manufacturing, but graphics and animation are generally underutilized in sales, management and other "non-technical" programming. Speedy Alka-Seltzer and the Pillsbury Dough Boy were examples of commercial animations that became the focus of entire marketing campaigns for basic consumer products. Similarly, a popular management skills videotape uses a "monkey on the back" graphic to denote situations where managers are lulled into accepting responsibilities that should have been delegated. Instead of keeping the monkey off their backs, these managers incorrectly take on accountability for tasks that belong to subordinates and co-workers.

CUE FADING TECHNIQUE

Purpose

To gradually enable the trainee to become less dependent upon verbal and/or visual explanations or "pointers."

Description

Cue fading is accomplished by gradually reducing the salience or "intensity" of instruction. It's like taking the training wheels off a bike—or starting with an inner tube, working your way to water wings, and then swimming unaided. Because video can explain concepts so clearly, and can distort reality in doing so, often we make the mistake of providing too many crutches for the viewer. While techniques like highlighted graphics, superimposition of words and arrows, and explicit narration are certainly justified when first treating a concept, when overused, they lead to problems in what's commonly called "transfer of learning." Why can the trainee understand the video program perfectly well—and perhaps even answer questions about it correctly—but fail to apply those concepts in the "real world"? A reason may be that things aren't quite as obvious in life as the talented video producer has made them.

Cues can range from blatant to almost subliminal. For instance, in explaining the parts of an engine, a program might start out highlighting the outline of each part, one at a time, in a specific color and labelling the part in that same color. Gradually, it might eliminate the label, and later, eliminate the highlighting altogether. Another way to cue trainees is by selective camera angles; for instance, parts of an item can be made apparent by selecting close-ups and composition where the trainee can't see anything but the portion under discussion. As the program proceeds, these cues can be faded by selecting shots more closely simulating a "normal" point of view.

As the trainee gains more experience with a subject during the course of a program, cues should gradually be faded so that more energy is needed to actually detect the cue. This can mean varying the size, color, or intensity of a cue so that it more nearly fades into the background. Of course, this must be done carefully so that it doesn't appear to be a mistake or just poor videographic skill. The information can be presented so that if viewers really need the support of a hint, it is available, but it also encourages them to function more independently and not always rely on the program to make critical decisions for them. For those who no longer need the support of the cue, it can easily be ignored and therefore the program can appear to be less patronizing.

Example

In any situation in which trainees must be taught to recognize behaviors or items, cue fading can be an ideal technique. For example, in training quality control inspectors, a program might start out highlighting defective parts with arrows, labels, close-ups and specific narration. As the program proceeds, these cues can be faded so that by the end of the program, trainees are seeing the parts as they will on the job.

Another interesting application of cue fading uses camera angles to train hairdressers. A large hair care and beauty products company distributes a training program to retail buyers who make volume purchases from new lines of hair care products. The camera frames a model hairdresser's hands at work in close-up, demonstrating specific techniques to use when applying the various hair care products to customers with different types of hair. The voices of using *verbal cues* to highlight the techniques and product features. The commentary becomes less didactic as the program progresses, concentrating on the "total look" as opposed to specific techniques. Yet, the visuals continue to depict close-up views of specific hair care techniques, which the audience is expected to observe and learn without detailed verbal explanation. The program ends with coiffured customers at leisure, smiling with friends, accepting compliments, and describing the benefits of their new hair care regimen.

REPETITION TECHNIQUE

Purpose

To show the same program segment repeatedly in either the identical format or from a different perspective in order to increase the viewer's exposure to selected information.

Increasingly, corporate/industrial video is structured in much the same way as commercial television. It is shot in real time using a traditional format (e.g., news, drama, comdey) with a beginning, middle and end. The Repetition Technique interrupts normal program flow to show the same material twice, and perhaps three times, when the level of activity, detail, complexity or importance merits additional emphasis. Usually, this "indoctrination" in a video segment is designed to enhance skill development, audience understanding of performance requirements, or the relative significance of an event or act.

Description

In its most basic form, the Repetition Technique is simply the presentation of a scene more than once during a program. The presentations are identical and may occur in succession or at different points in the program. In either case, the viewer is exposed to identical information, in every respect, two or more times.

This, of course, draws attention to the material and provides the viewer with additional opportunities to process information and retain it for future use. It is generally thought that two or perhaps three identical repetitions contribute to improved understanding and retention, while more than three identical repetitions leads the audience to see the material as excessively redundant and insignificant. Gimmicks can be used to alert the audience that it is about to see the identical scene. A character can call for a second showing by saying, "You can say that again!" Or a narrator might follow a particularly critical or complex scene with a request, "Let's watch Ellen again as she replaces that RAM board."

Information can be repeated in non-identical formats by manipulating presentation rates, camera perspective, sound and a host of other production variables. It is the creative application and sequencing of these production variables that draws attention, heightens interest, and enhances retention. If the audience has seen part of the program once, clever manipulation of production variables can "add value" to the second showing and make it even more informative and interesting. The aim is to *reveal* new information in the second showing which, up to that point in the program has been unavailable or understated.

Of the many production vehicles that can be used with the Repetition Technique, six of these have been selected and described below.

1. **Character perspective.** A scene can be repeated from a different perspective to provide the audience with an alternative view of the same set of circumstances. This second vantage point expands the viewer's information base by refining his/her understanding of a situation and how it could be interpreted by different observers. In a program where a sales team attempts to convince members of a purchasing department to place a sales order, two perspectives of the same meeting can be very different. Viewed from the perspective of the senior sales rep, the purchasing agents appear unsure and cautious. Shown a second time, but from the perspective of the agent, the vendors appear practiced and calculating. In sales, where "reading" the customer is essential to success, this second perspective is illuminating and important. From sincere and careful buyer to experienced bargain hunter, these two views of the purchasing agents remind the sales rep audience to balance first impressions with a detached view of customer motives.

2. **Camera lens and depth of field.** Switching between a normal, wide and narrow angle lens can broaden or concentrate audience attention. A wide angle lens distorts an image and provides great depth of field. The audience enjoys a wide field of vision, where movement to and from the camera is exaggerated and easily observed. A narrow angle lens is restrictive in depth of field and viewpoint as it focuses attention on a single subject or area of interest. Successive showings of the same scene, first with a wide angle lens and then with a narrow angle lens (or vice versa) provides an audience with access to new information.

3. **Camera focus.** The normal and near-normal angle lenses are particularly interesting in their own right when shifts in focus force the viewer to concentrate intermittently on foreground and background activity. The scene may have been shown initially with only the foreground in focus; a subsequent version of the same scene shifts the focus to the background where subtle movements take on greater significance. Focus can be "thrown" (background) or "pulled" (foreground) during a repeated segment to intermittently highlight information and direct viewer attention.

4. **Camera movement.** Once a scene is edited, the sequence of camera movements functions as a major vehicle in transporting the audience through the program. This is a linear pattern of camera movements over which the audience has no control. Repeating the same content using different camera movements can essentially offer the viewer a "second opinion" on what they've seen. A scene shot with relatively little camera movement positions the viewer as an observer, waiting for the performers to move and initiate action.The same scene shot with an active camera draws the viewer into the program as "participant." This subjective perspective encourages the viewer to decide why the perspective is shifting and what might be encountered as a result of the movement. (Although a zoom lens does not produce the same effect as camera movement, it can be used to approximate it.) As long as these movements do not distort the original script, set arrangements or program timing, they can persuade the viewer to take a more active part in the program.

A training program for bank tellers repeats a segment on the operational requirements for opening a home equity account. In the first showing of the segment, the audience sees a two-shot (teller and customer) and then pans with the teller as she gathers the necessary forms for processing. In the repeat version, the audience adapts the teller's perspective as the camera "walks" with her to the file cabinet to collect the necessary forms. In both versions, the teller walks to the file cabinet and verbalizes the name of each form as she pulls it. It is the subjective camera movement in the second showing that personalizes and reinforces the act itself as part of the customer service process.

5. **Sound manipulation.** A program segment takes on new life when it is shown a second time with no sound, selected sound, replacement sounce, or exaggerated sound. Any one of these tools can be used to heighten interest and improve retention.

No sound eliminates the need for the audience to concentrate on a second sensory channel. The message is less cluttered and visual detail is more pronounced. For example, a repeat showing of a program on downhill skiing entirely avoids off-screen narration. Instead, it relies on subjective camera movement and a collection of dramatic close-up shots to isolate key skill proficiencies. The audience of intermediate-level skiers has seen the program fully narrated and is now positioned to concentrate only on the visual message. The silence heightens the viewer's sense of isolation and excitement, while clearly exhibiting the required form and movement.

Selected sound removes some of the sound in a scene but leaves other sounds intact. The decision to exclude or retain sound requires an analysis of each unit of sound in the segment. A unit here is defined as a sound with a logical beginning and end such as a performer's line during an exchange of dialogue. When a sound is retained, it should improve the viewer's ability to comprehend the message and retain it for future use. For example, an emergency drill scenario is shown twice. The first is a standard version with background, performer and narrator sound. The second showing eliminates sound except when there is a direct reference to a specific emergency procedure. The sounds relating to procedures stand out against the otherwise silent movement of people, objects and camera position. The audience sees two successive versions of emergency drill operations, but in the second version, has its attention drawn by sound to specific action steps.

Replacement sound permits a producer to remove one sound in favor of another in order to achieve a desired effect. There are artificial or substitute sounds which can entertain as they gain attention and communicate a message. A courier driver is under pressure to make all stops on time without error. In the first showing of a program on driver safety, an Indianapolis 500 type announcer lends an element of suspense and excitement to otherwise everyday driving situations, e.g., "Now Ted is approaching a four-way stop . . . who has the right of way? The station wagon is still moving . . .". In the second showing, the play-by-play announcing is replaced by narrator analysis and decision making advice.

Exaggerated sound distorts the original sound by manipulating volume, speed, pitch, etc. A program in labor negotiations might include some do's and don't's for an audience of prospective negotiators. The program is shown once in traditional linear format. Shown a second time, each *key* negotiating skill is exaggerated by increasing the volume level when a performer identifies and explains it. This is a noticeable increase in volume, which draws attention to the important skills and fixes them in the viewer's mind.

6. **Timing effects.** These include slow motion, fast motion, freeze frame, graphic replacement or reverse motion. Assuming that one showing of a segment is in real time, a subsequent showing can use any of these controls to exaggerate and emphasize program content.

Slow motion exaggerates movement and increases the opportunity to study it. *Fast motion* adds a comic effect and speeds up an otherwise lengthy process. *Freeze frame* stops motion to emphasize a particular stage of development or moment in time. *Graphic replacement* substitutes an animated scene for a non-animated one to entertain, reinforce stereotypes or manipulate action artificially (e.g., an angry customer blows steam out of his head). *Reverse motion* repeats action in reverse to reinforce patterns of behavior, add comic relief or lend a mystical quality to a performer or place.

Example

A program on the subject of domestic pet care includes a series of vignettes, each shown twice. The first showing pictures pet owners in various situations failing to exercise proper precautions in caring for their pets. In the second showing, each situation is replayed with the pet's subconscious thoughts vocalized by off-screen narrators. Typically, the pet expresses its feelings in sarcastic terms with just enough seriousness to make the point.

For instance, in the first showing of a vignette, an owner fails to leash his large dog and seems unconcerned about potential hazards. The audience sees the owner reading a newspaper in a public park with the dog pacing in the background. In the second showing, the dog glances toward the owner and says to itself and the viewer, "So he thinks this is fun, no ball, no bone, no attention . . . I suppose I'll have to bring him a stick to play with . . . or should I growl and show my teeth to the Nelson's three-year old?" The camera shifts to background focus, revealing the three-year old playing with his toys. The sound of the dog's vicious growl sets the stage for a confrontation of dog, child and owner.

In this example, the first showing documents the owner's preference to relax instead of playing with the animal. The second showing adds new information regarding the pet's needs and potential danger in his owner becoming too inattentive.

DELIBERATE DEGRADATION OF VIDEO AND/OR AUDIO TECHNIQUE

Purpose

To establish a tone of fantasy, or to reduce the salience of the channel that is degraded and thereby divert attention to other aspects of the scene.

Description

While videographers usually attempt to make images and sound the highest fidelity possible, reducing that fidelity can achieve certain positive results in training. For instance, it is often necessary to separate a "fantasy story line" from real information; the fantasy sequences can be shot through filters which fog or distort the image. This is often used to represent changes in time or space, changes in an individual's behavior, or scenes that a character is re-living or imagining. Often, portions of a scene can be shot deliberately out of focus to call attention to the objects in focus while providing a context. To lead the viewer to shift focus from one object to another, the videographer can also literally change the focus, also called a "rack" focus. A similar effect can also be accomplished by reducing the luminescence of a portion of the picture, making it appear darker than natural.

Another form of degrading the video portion is to deliberately distort the camera movements or editing to make it appear choppy or unsteady. These techniques are used to heighten a mood of excitement, confusion, or drama. For instance, in shooting a mock confrontation between union organizers and managers, the videographer may shoot with a hand-held camera in the midst of the confrontation. This can simulate the subjective viewpoint, or can give the piece a "live news" feel.

Since virtually all video is now in color, deliberately changing to black-and-white can also add a dramatic effect. This can be used to simulate old film footage when doing historical pieces, or black and white photographs. Like the other distortion techniques, it can be used to signal a change of time or place, or another "side" of a person (see the Alter Ego Technique).

Taking video distortion to its extreme, it may even be effective to use a totally black screen while an audio track plays. This unexpected technique can gain the viewer's attention and focus it on the sound track.

The audio portion of a program can also be distorted for the same purpose as visual distortion. This can be done by passing the audio track through certain filters or even by editing the audio track or playing it backwards. For instance, in teaching the concept of tone of voice in a program on customer relations, the audio can be distorted so that the actual words are unclear, but the pitch and meter of the voice remain. This will make the point that how something is said carries as important a message as what is said. In other situations, it may be desirable to point out only the visual aspects of a scene because it is so highly loaded in verbal content. An example would be teaching about body language in interviewing. The actor's voice can be faded out or distorted through filters to focus audience attention on the ways characters communicate through body language, regardless of what they happen to be saying.

Example

In a program on phone sales for bankers, short sequences are shown from the point of view of the banker making the call. Immediately following, the same scene is shown from the customer's point of view, and the banker's words are distorted so that only the tone of voice and the pacing of the words is perceived. The emphasis is on total verbal delivery during a phone sale, not simply saying the right words.

A short program produced by Union Carbide to describe the counseling available within the company to employees having alcohol-related difficulties uses black and white images interspersed with color. The program uses one actor or a phone call reacting as if he were an alcoholic employee being informed about the counseling service. His "normal" self politely acknowledges the existence of this service, and then we see his unexpressed reaction of rage and frustration. The camera angle is distorted, and the picture turns to high-contrast black and white as he rants and raves about the fact that "sure, all I have to do is pick up a little phone and spill my guts . . . ." The black and white and distorted angle enhances the Dr. Jekyll and Mr. Hyde characterization and provides a powerful message about how this employee is probably feeling.

DELIBERATE OVERSTATEMENT (EXAGGERATION) TECHNIQUE

Purpose

To engage the audience in humorous overstatements of people and events in order to sustain interest and comprehension. By overemphasizing movement, effort, voice, reaction, size, etc., in ways that are funny or ridiculous, the audience is presented with a message that is at once unmistakable and enjoyable.

Description

The Deliberate Overstatement or Exaggeration Technique starts with a believable situation that has the potential to become tragic or catastrophic. The greater the potential for tragedy, the greater the leeway for producing a comic exaggeration of circumstances. It's difficult to effectively overstate a situation where the repercussions for flawed judgment, behavior or performance are insignificant. But, even a relatively tame concept can be successfully overstated if the program *context* has significance and tragic potential for the viewing audience. If Sally is eating dinner alone, there is little we can do to exaggerate her circumstances in a meaningful way. Introduce a mother who is obsessed with seeing her daughter lose weight through a "fad" diet to an audience of weight-watching traditionalists, and we open up a world of opportunity to deliberately overstate aspects of the food and Sally's reaction to both it and her mother. Here are some suggested guidelines for using the Deliberate Overstatement Technique to simultaneously communicate humorous circumstances and a clear message:

1. Write dialogue that is believable. The motivation for events or character antics must make sense to the audience. This is a gray area, because there are no rules to follow in marking the line between believable and funny vs. silly and boring. Good up-front analysis of audience attributes is required in order to take advantage of every inch of viewer tolerance towards a program's believability. Basically, the question is "This could be hilarious and it makes the point, but will they believe it?" If the answer is "No," it's a good idea to rewrite.

2. In general, target the exaggeration to contemporary themes. A treatment that portrays characters as "Star Wars" sound-alikes is more likely to strike a note with an audience than a reference to

the Dark Ages, which may be less familiar. Some story lines can work using early period themes or a mixture of themes from different periods. To be safe, test the proposed treatment against alternative contemporary themes before making a final decision.

3. Consider presenting a real situation in a totally abstract location, i.e., a glamorous fashion show in the streets of Harlem. The inherent exaggeration in contrast between event and location is readily apparent. This contrast acts as a natural backdrop for exaggerating specific scenes and corresponding messages.

4. With respect to shot selection, slight visual distortion (i.e., exaggeration) is recommended. Close-ups are useful for exaggerating facial reactions and selected movements. For example, when the harried sales rep can't get the copier to work during a product demonstration, his kicking the back of the machine is captured in close-up to exaggerate the frustration of the moment.

5. Regarding lens selection, slight wide angles add minor distortion to a character's face, making him/her appear more vulnerable. This draws audience attention and contributes to a subtle pattern of deliberate visual overstatement, where everyone or everything appears somewhat "bigger than life."

6. Camera movement presents an additional opportunity to exaggerate visual screen action. Much of the sub-plot in dramatic overstatements involves the individual's struggle to master, understand or overcome situational factors. By tracking action around these characters, either progressively from side-to-side or by actually circling a focal point, the intensity of the struggle to contend with circumstances is exaggerated. Tenseness that would otherwise be carried only by quick cuts and dialogue is heightened through this constant but focused movement, which gives the viewer a sense of limited time and space. The audience shares the character's feeling of helplessness or ineptness, and can still laugh at the humor in the situation. There is some risk with this type of camera movement in that it is also effective as an intensity-builder for non-humorous situations. So it must be used judiciously and in conjunction with other humorous devices.

7. Subtle effects can also be used to deliberately overstate the significance and impact of events. These "interruptions" distract characters and focus audience attention on the humorous circum-

stances that might otherwise go unnoticed. For example, a businessman is alone in the office late and remembers that an important report must be completed that evening. His only problem is inexperience with various pieces of office equipment. As he works frantically against the ever-present clock, his coffee spills, a printer ribbon jams, the cleaning lady enters sloshing water in her pail, the telephone rings, etc. Each of these distractions is magnified by exaggerating sound and camera shot selection. These effects are followed by character sighs, hapless expressions and periodic outbursts. All of this is funny because the audience has firsthand experience with similar circumstances.

8. Undoubtedly, the most important factor in exaggerating a situation is to begin with a cleverly written script and quality talent. The dialogue (or monologue) must take characters and events to the edge of believability, creating opportunities to exaggerate circumstances and make the points that need to be made. Variations in voice tempo, volume and pitch need to be built into the script. Shifts in body motion, i.e., quick jerks of the head, doubletakes, jumping unexpectedly, are necessary as well to maximize talent's capacity to overstate its reactions and attitudes. A good script demands good talent who can carry the exaggeration and walk the fine line between reality and fantasy. An arsenal of exaggerated expressions and attitudes are necessary for talent to qualify for this type of assignment.

Example

In General Electric's production, "The Critical Moment: How to Give an Effective Presentation," Salesman Miller is shown making both an effective and ineffective presentation. While the model presentation is impressive, it is also believable. The audience sees Miller as a clever and well-prepared sales rep who understands how to give an effective presentation. The botched presentation, on the other hand, is deliberately exaggerated for effect but it too is believable. Again, Miller appears to approach his assignment with all seriousness, but soon proves more inept than the viewer might think possible. Nevertheless, because Miller is making his product presentation to an ordinary group of sales trainees, there is enough credibility in the exaggerated segments to sound a clear warning to the viewer who thinks "it could never happen to me." In fact, it does happen and can happen to anyone who fails to understand and practice key presentation skills.

In any early segment of "The Critical Moment," Miller is first shown beginning a presentation the wrong way and then depicted doing it correctly. Like the other program segments, "beginning the presentation" is isolated and reviewed in some detail by a program spokesman.

An artist's drawing of still frames from "The Critical Moment" are shown in Figures 14 and 15 along with the accompanying script. (Still frames courtesy Penfield Productions, Ltd. in Agawam, Massachusetts.)

Figure 14. From "The Critical Moment," the **wrong** *way to begin a presentation.*

Miller:	"Oh, Gee! If one of you could just give me a hand. Oh, wow, that's quite a scratch there (MILLER CLUMSILY DROPS AND DAMAGES A PORTABLE OVEN PROD–UCT)."

Spokesman: "Does that sound like an effective way to start a sales training session on a new product line? Not if you want to capture your listener's attention. No. A fumbling disorganized opening immediately warns your audience that they're about to be bored."

Figure 15. From "The Critical Moment," the **right** *way to begin a presentation.*

Miller: (SALES REP DEFTLY INSERTS A TYPICAL 9" BAKING PAN IN GE AND COMPETITIVE OVENS. DEMONSTRATES THAT THE GE MODEL CLOSES SNUGLY WHILE THE COMPETITIVE PRODUCT DOESN'T.)

"Selling counter-top ovens is a game of inches." (DELIVERED WITH OBVIOUS SELF-SATISFACTION AS DOOR ON THE GE MODEL IS CLOSED.)

Spokesman: "That's the way to start an effective presentation. Miller's audience is hooked."

The same trainees listening to Miller are noticeably confused and embarrassed for Miller during his awkward opening and clearly impressed after his effective one. The point is that the model performance is noteworthy because it is contrasted with the deliberately exaggerated, almost slapstick, failed presentation. The latter is effective because it borders on the tragic and is embedded in the viewer's experience as having been an observer or participant in boring, ineffective presentations. Both segments are believable.

A second example of deliberate overstatement or exaggeration is too well-executed to omit. It is taken from the Avon production "Three Steps to Successful Selling." Mark Chernichaw, Avon's executive producer of television production, wrote about the technique in *Video Manager* (June 1984, 13). Here is an excerpt from that program where the "ideal customer" is exaggerated for effect.

Representative: Hi Jean . . . I'd like to show you so many great products and specials today.

Jean: I know . . . I know. I just can't wait to spend money. My husband Bill will be thrilled. I know it!

Representative: Great. Y'know, you look teriffic. Lose weight or something?

Jean: Oh, Helen, it's not me. I always look and feel so much better the day I know you're coming with your fabulous Avon products.

Representative You're a wonderful human being . . . (FREEZE FRAME).

Narrator (VO): If only every customer were the ideal customer. If only . . . but let's face it . . . life isn't always perfect.

(REVERSE TAPE ACTION AND HEAR A REWIND SOUND. SCENE BEGINS AGAIN. BUT NOW THE SUN ISN'T SHINING. HOUSES LOOK A BIT DRAB AND LAWNS AREN'T SO NEATLY MOWN.)

In this production, it is primarily verbal exaggeration that makes the audience laugh. While it is unlikely that such an ideal customer can be found in the real world of selling, every sales rep has encountered prospects who are more interested in buying than others. As the program addresses the problem of unwilling customers and how to overcome obstacles and objections, the exaggerated image of the willing buyer remains. It is an image that cannot be entirely extinguished because every experienced sales rep has sold to that rare someone who has made selling easy and profitable.

DRAMATIC IRONY TECHNIQUE

Purpose

To steer an audience through a set of circumstances to unexpected (sometimes opposite) conclusions. By permitting the audience to understand the incongruity between events and outcomes, while characters in the program remain unaware, the real significance in a message can be both humorous and compelling.

Description

When irony is used to communicate, the trick is to show an audience that things aren't always what they seem. To the extent that the audience can laugh at the incongruity between what is and what should be, the program will command attention. With that level of interest comes a better understanding of why the circumstances are funny and how their message relates to audience needs, goals and behavior.

In producing dramatic irony, be on guard against creating a put-on. The viewer will truly accept the situation only if it is real and honest. The viewer must interpret the program as a legitimate representation of self or others. Once the line between realism and fantasy is crossed, the audience is quick to reclassify the concept as off-target, unrealistic and inapplicable. Maximizing the value of humor while maintaining realism—that's the challenge in making dramatic irony work.

Another consideration when using irony is to make sure the punch line is on time. Under no circumstances should a punch line appear separate from or in addition to the situation it addresses. Often, programs with ironic circumstances require dialogue that runs the risk of sounding like it was written to "get a laugh" instead of being an integral part of an entire scene. The issue is *comic timing.* Remember, irony is intended to elicit humor because it contrasts unexpected outcomes with believable circumstances. Humorous punch lines that linger signal the audience that the circumstances are fabricated. While the situation may still be amusing, the audience cannot be expected to seriously apply what they've seen to everyday life. Dramatic irony isn't tacked onto the situation—it *is* the situation.

With this technique, the writer, director and talent have a common objective: to sequence the program so that the audience experiences a *shock of recognition* when the ironic turn of events takes place. To meet this objective, writing and acting are interwoven to ensure that dramatic style is maintained. Once this consistency is established, a sort of "quality control" emerges to reduce the risk of separating the humor from the message. The audience is positioned for the unexpected and the situation becomes that much funnier, even ridiculous, when they recognize that "it can and does happen."

Here are five guidelines to follow in producing dramatic irony:

1. Employ talent with a repertoire of facial expressions and head movements. These are important in emphasizing the responses characters have towards ironic events. If not overused, reaction shots can be made more engaging by having talent turn toward the cameras to share surprise, confusion or amusement with the viewer. The general message for the viewer is "Did you see what I saw?" or "Can you believe this?"

2. Before writing a single word of script, determine the ironic turn of events and how they will be portrayed. Humor in general, and irony in particular, requires such sensitivity towards the fine distinction between what is funny and what is not that every scripted word and action must contribute to *making the irony work*. Once the irony of the situation is conceptually linked to the intended message(s), the production team can use it as an objective in structuring the program.

3. Limit the use of irony to a single major outcome, with perhaps a few minor, clearly related interpretations. The audience may suspect but should not be able to conclude that the chain of events in a program will lead to an ironic twist. There should be no attempt to "clue" the viewer into the likelihood that all is not what it seems. Content should be substantively related to program objectives, using irony as the vehicle to drive the message home precisely because it is totally unexpected.

4. Music is especially effective in accenting the existence and effects of irony. A light, airy tune can accompany an expression or group reaction to an ironic outcome that is unexpected but not particularly harmful. A "Dragnet" theme or similar foreboding

musical cadence can signal a turn of events with serious, yet comical, overtones.

5. Spend more time than normal blocking out scenes with ironic implications. Experienced talent can usually offer a range of interpretations in addition to those originally intended. The unrehearsed walk-through of a scene's possibilities takes the production team into variations on sub-text that are often ignored when the shoot is live. At times, this type of exercise may take a program beyond its capacity for humor, but the director must be prepared to explore these subtleties and ready the production for its most potent interpretation.

Example

Here is an early version of a treatment for a program on basic sales strategy. The organization is a major competitor in the office systems industry. The concept is to employ irony to contrast effective and ineffective behavior for sales representatives who use entertainment or business meals as sales vehicles. There is plenty of opportunity here for humor, as Larry (the sales rep) is unaware of the competitive sales presentation at the next table. He is so pleased with himself and so intent on talking instead of listening that he notices little else. His sales strategy is off-the-mark and inappropriate given his client's needs and concerns. By contrast, the conversation at the neighboring table is precise and tactical. Larry seems comical because he plays it straight, pointing out the wrong way to conduct this sort of sales call. The audience must appreciate Larry for what he is (well-meaning, sincere and far too typical) and isn't (polished and alert), never seeing him as a comedian, but rather the kind of salesperson who lives by style and lacks substance. Larry is the kind of sales rep we hope never to become. While he makes us laugh, it is the realism of his behavior that makes the message stick.

Concept: "Larry's Favorite Restaurant"

When Larry, the ABC sales rep, takes an attractive female prospect to the finest restaurant in town, touting it as "the home of his greatest business deals," he sets the stage for an ironic turn of events. Once his overconfidence and skill deficiencies are established early in the program, Larry's prospect begins to signal her objections to several of his assumptions. Larry is blind to these obstacles and continues to comment periodically on the attributes of his "favorite restaurant."

As Larry delivers his pitch, the prospect begins to pay attention to a sales presentation (at the next table) by a representative of Larry's chief competitor. Although she remains superficially involved in conversation with Larry, the prospect responds to what she hears by visualizing the competitor's product at work in her work environment. The audience shares in these short dream sequences which are interrupted periodically by Larry's insistent voice and annoying wrist-grabbing.

In the end, the restaurant which held such promise for the sales rep at the start of the call turns out to work against him. A trip to the men's room and Larry returns to find his prospect leaning into a three-way conversation with the gentlemen at the next table. As he approaches, the audience hears Larry's would-be client make arrangements for a product demonstration in her office. The prospects for Larry's competition to make the sale are good.

Larry, on the other hand, never realizes that he selected a table only a few feet from his competition. He assumes someone struck up a conversation in his absence and is, in fact, pleased to meet the competitor. On the way out of the restaurant, Larry learns that his client has no plans to pursue his product any further. She does thank Larry for lunch, suggests that they do it again and agrees that "this really is a fantastic restaurant for doing business." With this remark and Larry's dumbfounded expression, the irony of the encounter is sealed and Larry's dull sales approach and false confidence are verified.

3

Interactive Design Techniques

Although interactive video is actually quite a separate technology from traditional linear video, the techniques for both can be thought of as being on a continuum rather than in dichotomous categories. Good instruction should always encourage activity in the learner, whether overt or purely mental; however, the conditioning of many people from broadcast TV leads them to become quite passive in front of the "tube." Interactive video was invented to overcome that passivity and to enable programs to be tailored to the individual. Through the use of playback controllers and/or "branching" programs, training can be made more personal and adaptable to individual skill and knowledge levels, interests and backgrounds.

Interactive video is too often defined by its hardware (i.e., a videodisc tied into a microcomputer), whereas it is more helpful to approach it on the basis of design techniques. A program viewed in a continuous manner on an ordinary VCR might be highly interactive in that viewers are drawn into the program and actively participate by taking notes, modeling demonstrations, or coming up with examples and applications in their own heads. On the other hand, a program displayed on a sophisticated simulation system capable of multiple inputs and branching is not interactive at all unless someone actually participates in using it—both physically *and* mentally.

Because interactive video can take many forms, it's best defined as *programs which require viewer response*. Typically, systems are composed of a video playback device (tape or disc) controlled by some external microprocessor like a keypad device or computer. However, interactive techniques can be supported using standard playback hardware without hooking up peripheral controllers.

Figure 16 presents a chart of the "Levels of Interactivity" developed by OmniCom Associates to assist individuals in categorizing and selecting interactive hardware/software systems. It progresses from simple techniques using ordinary equipment to highly specialized programming and systems. (Note that this classification system is different than the "Levels of *Videodisc*" taxonomy often used to describe whether videodiscs are controlled by internal or external computers.)

Level one, "direct address," is a technique which should be used in almost every type of video script, whether it's thought of as typically interactive or not. The writer simply uses the second person and talks directly to the viewer as if it were a one-on-one conversation. Rhetorical questions can be asked to encourage viewers to answer "in their heads," and other "closure" techniques can prompt trainees to complete the picture or sentence vicariously. (Remember the famous ad, "Winston tastes good like a . . . . ?") Of course, you can't tell whether trainees are actually participating in the program actively, but simple writing formats like this one can make any program a bit more interactive.

Level two, "pause," uses the ordinary controls on a video tape or disc player to allow trainees to control the rate, direction, or order of a program. Many organizations use video with accompanying workbooks or lab exercises so that the video becomes the stimulus that requires some overt response from the trainee. (See the description of the "pause" technique for more details.)

Level three marks the first appearance of "branching," or the ability of a program to display different segments to different people depending on their responses to questions. This level is characterized by the use of simple random-access controllers which can fast-forward or reverse a tape or disc to a desired place either by counting control track pulses or by accessing individual frame numbers on a disc. These controllers are inexpensive and are both the programming device and the student response device. For tape use, they work with random-access capable players or recorders which are capable of counting control pulses when in the fast-forward or rewind mode, and which have

	PROGRAM DESIGN	HARDWARE	QUESTIONS	DATA COLLECTION	AUTHORING
"INTELLIGENT" SYSTEM	recursive	specialized	natural language comprehended	data modifies program	specialized authoring/ programming language
RESPONSE PERIPHERAL	branching		motor responses evaluated	responses can be recorded and summarized	specialized programming
MICROCOMPUTER			constructed answers evaluated		authoring system/ language
RESPONDING DEVICE				choice and latency recorded	authoring device
RANDOM ACCESS			multiple choice with feedback	none	read/write controller
PAUSE	linear	traditional	self-evaluation		none
DIRECT ADDRESS			rhetorical		

Figure 16. Levels of interactivity.

the ability to be remotely controlled. In this level, questions are presented within the video program, and students answer by selecting a specified segment, frame, or chapter number as their answer. For instance, to answer a question with this level of hardware, the trainee would need to press a sequence like "C25S"; "C" to clear the previous response, "25" to call up segment 25, and "S" to initiate the search. Branching allows you to ask questions and give individual and immediate feedback informing the trainee about the correctness of the response, and if incorrect, to provide further explanation. This segment-accessing ability also allows one to include "menus" from which viewers can select examples or the order in which they'd like to experience the instruction.

Level four uses dedicated microprocessors specifically designed to facilitate interactive video with tape or disc. In this level, questions are presented on the tape or disc, or are stored in a digital form and generated into "screens" of text which appear on the monitor. The codes to control branching are stored right on the tape or disc rather than on a separate microcomputer floppy disc. Students answer questions by pressing a number on the tape or disc controller; here, they can simply key in one number rather than the more lengthy segment number as required in level three. Some of these systems can provide a print-out of trainee responses in terms of the correctness of the response, the score, and the length of time it took them to make each response. Level four requires more expensive playback hardware and also separate programming hardware. These systems are preferred over level three where more branches are needed (most level three controllers only access up to 64 segments whereas level four systems can usually support about 200), where printouts of responses are desirable, and where the keying in of responses needs to be simplified. This configuration when applied to videodisc, is often referred to as a "Level II" System.

Level five is what most people think of as interactive video—the interfacing of a microcomputer with a tape or disc. This level allows for more flexible programming and responding, but is also more expensive and difficult to produce. For these systems, you need not only the microcomputer and video player, but also an interface which enables them to "talk" to each other. Programming of the computer which presents text screens, accepts and evaluates responses, and controls the playback of the video is done either by writing programs in a language like BASIC, or can be facilitated by various authoring aids without the need to learn the code and syntax of a language; features are selected from menus written in ordinary English and text screens are written in the way that they will appear to the end-user. Authoring systems are easy to learn, but are less flexible than authoring languages in supporting a range of interactive techniques. Authoring languages are specially designed programming languages whose code is mnemonic to training applications, and whose structure is set up to facilitate the answer-judging routines that are necessary to support interactive video and computer-based-training. Authors must learn a foreign code and syntax to use authoring languages, but can create more varied and sophisticated programs with them than with most authoring systems.

Level five systems allow trainers to design programs which require constructed answers of trainees (fill-in rather than just multiple choice). These systems can also support calculations for simulations, display computer graphics, analyze and print-out response patterns for individuals or groups, and can allow segments of instruction to be presented by less time-consuming and costly text screens rather than merely by video. These hardware/software systems are more expensive than other levels, and to be fully utilized require more sophistication in lesson design and programming. For some applications, computers are less desirable than simple keypads, especially in cases where the computer might be easily damaged or where trainees are not adept at typing. This type of videodisc/computer system is also known as a "Level III" videodisc.

Level six adds a response peripheral to an interactive video system; these devices allow trainees to respond in an easier or more natural manner and can more closely recreate actual on-the-job equipment and situations. These systems can range from simple game controllers and touchscreens to sophisticated simulators. (See the description of the response peripheral technique for more details).

The highest and most abstract level is "intelligent systems"; these programs are *recursive* in that they actually learn and modify themselves in use. They aim to more closely recreate a natural communication system by their ability to respond to a range of natural-language responses and to actually improve their ability to "understand" trainee vocabulary and styles. Artificial intelligence is an important trend in training, both for understanding the ways in which people learn, and for creating systems which are better able to assist in the learning process.

Since it's quite clear that interactive programs usually cost more money to design, develop, and administer then simpler media, why is the technique so popular? The answer is because this technology provides for perhaps the most flexible training with the exception of individual human expert tutors for each student. The ability to branch and tailor programming to individuals and to present a range of sound, motion, color, text, and graphics makes interactive systems adaptable to a variety of training situations. If questions are designed well, you can ensure that trainees have mastered the subject matter; otherwise, they wouldn't have been able to finish a program! Print-outs of trainee progress can be kept in their files to document their skills for personnel evaluations and to document safety training for insurance purposes. Overall, interactive video can often supplant traditional classroom training, saving money and allowing the instruction to be provided in a more flexible manner. Studies seem to point to the fact that interactive programs can also save time, since trainees see only what they need to know, rather than having to sit through live or mediated presentations which must be aimed at the lowest common denominator.

Many of the design techniques for interactive video are the same as for linear video; novel approaches for dramatization and visualization can be used in either format. However, to produce good interactive video, you must be able to think "branchingly" and anticipate not only what the *right* answer is, but more importantly, what the likely mistakes will be. For each misconception, you'll need to come up with a branch which explains the concept in relevant terms. Designing a good interactive video program is more like planning an interview than writing a speech. You must not only concentrate on what *you'll* say, but on what to ask, what the person is likely to say in response, and what you'll say for each category of possible response. Effectively used, interactive video is probably our best shot at incorporating all the instructional technology at our disposal today.

PAUSE TECHNIQUE

Purpose

"Pause" is the basic level of interactive video which allows the trainer to integrate video information or examples with some other training activity, such as hands-on practice or discussion. The purpose of this technique is to permit viewers to control the pace of the playback as well as to add a dimension of activity to the process.

Description

The Pause Technique for interactivity can be employed without the need for specialized interactive hardware. Whenever a video playback device is available that can be stopped and/or paused by the user, the responsiveness of the presentation can be increased along with the activity level of the trainee. The Pause Technique simply means scripting an ordinary linear presentation so that the user can stop or pause the program to engage in some other activity. A typical example is the videotape/workbook training program where the video presents general concepts and the workbook is used to highlight information and provide individual exercises. The workbook might contain answers to the questions at the end, or, alternatively, that kind of feedback can be handled by the next portion of the tape. Typically, the video program is divided into short topical segments which end with the direction "Now stop the tape and ----- (go to workbook section X; do the first part of the assembly, etc.)"

Another application of the Pause Technique is to "branch" the trainee to some other activity. In a small group presentation, stopping places can be added in the program during which a class can engage in a discussion. Programs can also be constructed to accompany a lab activity or an exercise in learning to assemble objects.

A number of organizations use this method to teach line workers to assemble parts; trainees watch a short segment of a program, stop or pause the tape, and then model the behavior shown in the tape by following the same sequence of tasks. Of course, if the person needs to see the sequence again, he or she can replay the segment. This technique can be used to train new employees as well as experienced workers in assembling new devices. Organizations that have used this technique find that it speeds productivity and helps to maintain standardization of procedures.

In order to use this technique effectively, it must not only be possible for users to control the video playback device (whether videodisc or videotape) but also they must be ENCOURAGED to do so. Most of us have grown up in the tradition of broadcast TV where it wasn't possible to stop the presentation—except to turn it off and forever lose the possibility of seeing any more of it. Many people aren't comfortable with video equipment, and think that it's not their role to "play around" with the machinery. Sometimes, adding an inexpensive remote control with easy-to-read playback keys will make it easier for people to interact with the presentation instead of leaping out of their chairs to figure out which button should be pressed on the VCR or disc player. The video presentation may begin by showing how the particular model of video device used can be stopped and paused—that is, if there is any kind of standardization across models of playback hardware. A tendency is for people to ignore the stopping places and race right through the program, looking in the back of the workbook for the answers, or filling them in as they watch the follow-up segment on video. This can be avoided by calling for rather detailed answers to be filled in the workbooks which can be collected from each person for evaluation. In the case of the "pause" method being used to teach some procedural or psychomotor skills, one might warn the audience that what is being taught is complex and that they'll have to do it step-by-step in order not to get confused. In short, have in mind some good reason to employ stopping points in a program, or trainees will figure out a way to go through it more quickly. Since the use of this technique requires that the trainee to go along with directions, it can be a problem for recalcitrant viewers. However, if people are truly motivated to learn something, this technique can enhance trainees' participation in video learning and provide interactivity without the expense of any additional equipment.

Examples

"Pause" is most effectively used when presenting complicated instructions which need to be practiced one step at a time, when teaching some psychomotor skill, or when it's desirable to include group discussion to individualize a presentation or to encompass a variety of viewpoints. Make sure that the program gives the audience the "permission" and the confidence to manipulate the playback device, and that they understand the reasons for interrupting the video playback. This "basic" kind of interactive video is enhanced when the nonvideo activity is itself highly interactive.

Carrier Corporation uses a series of video/workbook exercises to teach supervisors to use their new computerized employee time and attendance system. The video program demonstrates each function, and then instructs the viewer to stop to tape and employ each function by using a terminal hooked into the system right next to the VCR. Trainees can access "dummy" employee files that can be manipulated according to the taped demonstrations and specific directions in the workbook. After completing each exercise, the trainee views the next tape segment which shows what the trainee's terminal should now look like, and what some frequent causes for error are.

The U.S. Bureau of Prisons has, since the early 70's, used the tape/workbook technique in its video training for correctional officers. Tapes were divided into segments and after viewing each, trainees stopped the tape and answered questions in the workbook. An answer key was usually provided in the back of the workbook so that responses could be checked.

A major utility uses a video/workbook package to train supervisors to write better performance appraisals. The videotape presents interviews with actual employees discussing techniques and real situations, and the workbook contains copies of actual performance appraisals. Trainees also watch dramatized scenes, write appraisals, and compare their documents with appraisals generated by experienced supervisors.

CONTROLLED PLAYBACK TECHNIQUE

Purpose

Video's ability to freeze, slow down, reverse, and replay events is useful in teaching skills and concepts involving motion. Often, events in the real world occur too quickly for the novice to attend and understand; by videotaping those events, trainees can control the pacing. Even when "live" training is available or being used, this built-in aspect of the medium can be put to good use since it can make concepts, especially those involving motion, more salient to the untrained eye.

Description

Controlled playback can occur with many late-model video cassette recorders, and even more elegantly, with videodisc. It's important to assess if the playback device will allow the viewer to pause, scan forwards and backwards, or play the tape in slow motion. Once it is determined that the VCR supports these functions, play a tape using them to see the quality of the video when it's not being played in the normal "forward" mode. Many of the newer players will roll the "noise bar" out of the picture during pause, and some also feature "dynamic tracking" which stabilizes the picture in fast or slow motion.

No unusual measures need to be taken during production of a program which uses this level of interactivity, except perhaps to include a suggestion to viewers in the beginning of the program that they utilize these controlled playback features. Note that this eliminates the need to expensively create slow motion scenes during editing; this effect is totally accomplished during playback.

Of course, most VCR's and videodiscs will mute or distort any audio whenever the program is not being viewed in normal speed playback, so don't plan on having any important audio track over a segment that will only be seen in slow or fast motion. However, a common technique to use is to show a shot in normal motion using voice-over or "live" narration, and at the end of the scene, tell the viewer to review the scene in slow (or fast) motion. The audio won't be heard the second time around, but that shouldn't be necessary, especially if the normal speed narration emphasizes key visual elements, i.e., "pay special attention to . . .".

This "controlled playback" feature is fully utilized in videodisc programs where still frames can easily be combined with regular and fast-motion scenes. While still frames using up just one frame of video would go by at 30 per second during normal playback, the viewer (or an outside computer control) can show them one at a time, or in slow motion. This can be done with videotape also, but keeping still frames up for longer than a few seconds starts to wear out the videotape; this wear does not occur with laser optical videodiscs, so still frames can be watched virtually indefinitely. Devices which can play back digitized audio over still frames can create minutes of "slide-tape"-type presentations using up just a few of the 54,000 frames on a videodisc; the audio is stored in a digital form in unused areas of the video signal. Where portions of a program don't absolutely need motion, these still-frame sequences allow you to pack a great deal of information on one disc. Likewise, fast-motion sequences can be stored on a disc which, when played back in slow motion, simulate normal speed; this technique can also be used to save space on a disc.

Example

OmniCom Associates has produced a videodisc on the sign language of the deaf. This program allows a hearing viewer to vicariously explore the world of sign language, and to "stare at" and study in freeze-frame or slow-motion actual sign sequences such as excerpts from real conversations, theatrical signing, interviews, signed stories and songs, and a sign dictionary.

INTRODUCTORY MENU TECHNIQUE

Purpose

The purpose of using a menu to introduce a training program is to allow viewers to select appropriate information, either based on their needs, interests, or a particular situation. A problem with most linear materials is that people are required to sit through material which doesn't apply to them; fast-forwarding or rewinding to particular segments can be awkward with most equipment, and trainees might not know which portions of the program to seek out, anyway. The overall purpose of using introductory menus is to save training time by letting users access the information they need when they need it. This efficiency also can add to trainee motivation since "wasted time" is reduced, if not eliminated.

Description

An introductory menu is most easily used with an interactive device capable of automatically selecting a segment based on a simple input. "Level three," "level four," and "level five" hardware/software systems are generally applied. However, it is quite possible to include introductory menus using no specialized interactive hardware

If playback equipment is standardized, users can be taught to use the built-in counter on the VCR by rewinding a tape, "zeroing out" the counter numbers, and then fast-forwarding to a certain counter number to see particular segments. If this is not possible, many videotape and videodisc players will fast-forward in the "scan" mode; segments can be identified with about one minute of a title over a distinguishing color background. Another approach is to "key" or superimpose a "chapter" number in the corner of the video screen throughout a section. Using these, viewers can scan until they see the appropriate color background or number whiz by, and then watch the segment. Of course, using color-coded scanning or counter numbers does not allow for very accurate accessing, but by leaving enough of a "pad" of a title, most viewers don't have trouble in using these simple devices.

Using random-access interactive equipment, of course, finding segments can be done much more accurately and easily. "Level three" devices allow you to enter a pre-defined segment number into a small keypad to search out a segment. The more sophisticated hardware allows for entering a number from a menu list, or even a word if a keyboard is used. When producing interactive programs for any of these levels, it's important not to leave viewers "hanging at the end of a branch"; be sure to specify how the viewer can end the program or access other segments.

Front-end menus are effectively used when trainees don't typically have to see an entire program or when the order of presentation isn't important. When trainees have varying levels of incoming experience, menus can let them skip over material they already know—IF they can judge this for themselves. Long programs can be broken down into individual segments allowing trainees to go through each part slowly, making reviews easier and lengthy viewing sessions unnecessary. These menus also aid re-training, since people can pick out the individual parts they need to review. Introductory menus can also be used to identify certain model numbers of equipment or steps in a process about which people might need information, or be used to identify regional or job-related differences in policies or procedures.

A note of caution here: this technique should not be used when each person needs to see all the training and/or when the material needs to be taught in a certain order. Menus give a sense of "power" which is good, but can tempt people to skip around and omit important portions, making their knowledge confused and incomplete. Don't ask trainees to decide what they need to see if they are not in a good position to judge that for themselves.

The menu approach, then, is akin to dividing a book into chapters: it can make training more efficient by letting people easily select segments, and it can be used to tailor presentations to individuals, situations, or regional variations.

Examples

The National Training Fund for the Sheet Metal and Air Conditioning Industry has produced a number of videodiscs to support live stand-up training through Robert Drucker and Co.'s production facilities. The instructor's manual which accompanies the set of discs contains chapter and frame numbers so that trainers can call up relevant examples and quizzes during classes. The discs themselves also begin with indices. Following is a section of a disc index. (See Figure 17.)

Introduction to Soldering

Soldering has historically been vitally important in the sheet metal industry. However, in recent years, other methods of joining metals and waterproofing have reduced the amount of soldering required in sheet metal work. But knowledge of soldering techniques is still essential for sheet metal workers, especially in the Architectural and Industrial Sheet Metal fields. The National Training Fund has produced this videodisc on Soldering to assist instructors in their apprentice and journeyman training programs.

This two-sided videodisc includes an overview that describes the importance of soldering in the sheet metal industry. It is followed by a module that outlines the equipment and material required for soldering. On the second side of the videodisc is a module that shows the proper soldering techniques and common problems for different metals of varying thicknesses. There are also two audiovisual work exercises that describe proper methods for different joints and positions.

The still frame appendixes include a list of the gages for different metals and multiple choice question quizzes. The following are the main and chapter indexes for this National Training Fund videodisc on Soldering.

PROCESS & EQUIPMENT AND MATERIALS (Side A) Disc #08084

INDEX OF CHAPTERS

Chapter	*Description*	*Frame*
1	Indexes for all chapters	1000
2	Gages of Different Metals	1100
3	Quiz	1200
4	Film: **Soldering in the Sheet Metal Industry**	2000
5	Film: **Soldering Equipment and Materials**	11500

Figure 17. Excerpt from videodisc menu, courtesy of the National Training Fund for the Sheet Metal and Air Conditioning Industry.

CHAPTER #2 INDEX—Gages of Different Metals—Frame 1100

Description	*Frame*
Sheet Steel	1101
Galvanized Steel	1104
Stainless Steel	1107
Aluminum	1110
Copper	1113

CHAPTER #3 Quiz—Frame 1200

The visual answers to the following multiple choice questions that are on this side of the disc can be found between the listed frame numbers.

Question	*Frame Numbers*
1. The number stamped on a soldering iron is the weight in pounds for: 1. **A pair of soldering irons.** 2. A single soldering iron.	13685-14170
2. You can clean a soldering iron tip by using: 1. A grinding wheel 2. **Files** 3. Muriatic Acid	15030-15590
3. 50/50 solder melts at about: 1. 200 degrees 2. **400 degrees** 3. 600 degrees	22980-23180
4. The most commonly used solder is made of: 1. Zinc 2. Terne 3. **Tin and lead**	22480-22670
5. Sal Ammoniac is used to: 1. Draw a soldering iron. 2. **Tin a soldering iron.** 3. Clean a soldering iron.	15660-16165

SOLDERING

Question	*Frame Numbers*
6. The proper color for the flame in a bench furnace or portable firepot is: 1. Yellow 2. Green 3. **Blue**	17624-18004
7. The word "flux" means: 1. To join 2. **To clean** 3. To flow	24400-24620
8. Non-corrosive fluxes are most commonly used on: 1. Stainless Steel 2. **Copper** 3. Galvanized Steel	28620-28840
9. Muriatic Acid can only be used as flux on: 1. Stainless Steel 2. Copper 3. **Galvanized Steel**	27640-27790
10. Capillary Action is sometimes referred to as: 1. **Sweating the solder** 2. Soft soldering	5310- 5600

CHAPTER #4 INDEX—Film: "Soldering in the Sheet Metal Industry" (4 minutes)—Frame 2000

Sequence	*Frame*
Start of Film	2061
Definition of Soldering	4630
Advantages of Soldering	6760

CHAPTER #5 INDEX—Film: "Soldering Equipment and Materials" (13 minutes)—Frame 11500

Sequence	*Frame*
Start of Film	11561
Types of Irons	12190
Tinning an Iron	14785
Drawing an Iron	18340
Bench Furnaces	19410
Portable Fire Pots	19970
Types of Solder	22440
Types of Fluxes	24267

Figure 17 (continued).

INTELLIGENT MANUAL TECHNIQUE

Purpose

Interactive video programs can use sound, motion, color, and branching to become an "intelligent" manual. This format is used to assist individuals in repairing and/or operating equipment and in diagnosing situations; the branching can be used as a tailored diagnostic aid to trouble-shoot machinery or processes and enable the user to quickly find and solve problems. This application of interactive video is usually more of a "job aid" than a training device for novices.

Description

Branching video manuals are usually patterned after the artificial intelligence model of "expert systems." They attempt to map and "can" the facts and procedures in an area so that other, less skilled persons can utilize the knowledge base to spread the "expertise" around. The content for such "expert systems" and intelligent manuals is usually identification or diagnosis and repair of some system—be it a machine, a human body, or a social/political situation such as military decision-making. First, a content area's diagnostic procedures are identified: this is usually done in terms of if/then statements used to gradually narrow down potential problem areas. Interviewing expert repairpersons will help develop a flow chart of questions which lead to diagnosing a problem and then taking care of it.

For example, if one were to develop an intelligent manual on how to cook a steak, one would want to include a section on how to tell whether it was done or not. First, the program might ask what color the steak was after it had been cooking for a few minutes. It might show a split screen of steaks in varying degrees of doneness and ask the trainee to match his or her steak to one of the pictures. If the color were bright red, the trainee could be asked to check the temperature of the steak. If it were still cold, the trainee could be told to check if the grill was hot. If the grill were cold, the trainee could be instructed to check to see if the power was turned on. If the power were on, the program would ask the trainee to see if the unit was plugged in and the electricity working. Obviously, the program would include graphic, pictorial, auditory, and text sections asking pertinent questions and providing instructions on what to do next for each contingency.

True expert systems for complex situations are more in the realm of simulation development than training, and may involve sophisticated research and mathematical models. However, if there are a modest number of questions and alternatives in learning to identify and solve a problem, this technique is not beyond the reach of trainers and video producers. If there are any "holes" in the system, however, it cannot be successful. Expert systems provide a structured approach to explaining diagnosis and remediation procedures and can be used for experienced employees or can be adopted for use with novices in laboratory set-ups.

Example

An example of this technique is a program produced by General Electric to diagnose problems with locomotive engines. The program was generated after interviewing GE's top service engineer over several months, and then breaking down his accumulated knowledge into a series of over 500 rules and facts about locomotives. Since there are few human experts on this topic, when this system is fully implemented, there will be a computer "expert" at each repair shop, making the work much more efficient. The system uses a powerful computer which can access still frames or motion sequences for repair demonstrations from a videodisc. The system first displays a menu of possible symptoms, and then proceeds through more detailed questions like "Is the fuel filter clean?" or "Are you able to set fuel pressure to 40 pounds per square inch?" Following is an example from a GE flowchart and a photo used in the program.

Figure 18. Photograph of GE repair engineers using intelligent manual or "expert system" to diagnose problems with locomotive engines. Photograph courtesy of General Electric Corporation.

Rule 760
there is a fault in the fuel system at idling speed and readings were taken from locomotive fuel pressure gage
IF:
EQ [ENGINE SET IDLE]
Is the engine at idle?
EQ [FUEL PRESSURE BELOW NORMAL]
Is the fuel pressure below normal? (Less than 38 psip)
EQ [FUEL-PRESSURE-GAGE USED IN TEST]
Did you use the locomotive gage?
EQ [FUEL-PRESSURE-GAGE STATUS OK]
Is locomotive gage known to be accurate?
THEN:
WRITE [FUEL SYSTEM FAULTY] 1.00
establishes that there is a fuel system fault.
End of rule 760

Rule 1270
the locomotive fuel-pressure gage is OK
IF:
UDO [FUEL-PRESSURE-TEST-GAGE STATUS ATTACHED]
Attach a known good pressure gage.
ASK-Y [FUEL-PRESSURE-TEST-GAGE READING SAME-AS FUEL-GAGE]
Is test-gage reading the same as locomotive-gage reading?
THEN:
DISPLAY [FUEL-PRESSURE-GAGE STATUS OK]
The locomotive-pressure-gage is OK.
WRITE [FUEL-PRESSURE-GAGE STATUS OK] 1.00
establishes that the locomotive-pressure-gage is OK.
WRITE [FUEL-PRESSURE-GAGE STATUS ALREADY TESTED] 1.00
establishes that the locomotive-gage has been tested.
End of rule 1270

Figure 19. Sample rules from GE's Expert System flowchart on diagnosing problems with locomotive engines.

VISUAL DATA BASE TECHNIQUE

Purpose

Interactive video can be used like an encyclopedia with motion, color pictures and sound. Using routines which access segments or still-frames by word, phrase, or number, these programs are useful as training references for independent inquiry or as an aid to stand-up training.

Description

A videotape or videodisc is produced to contain a number of short scenes and/or still frames. These can then be accessed simply through a random-access controller, a device which searches out a specific counter number on a tape, or by searching for a specific frame number on a videodisc. These numbers or segments can be indicated by an accompanying print document or for programs with a limited number of segments, on an introductory screen of the video. More options for which to search can be supported by using an external computer interfaced with the video player. A series of menus can be presented to the viewer, which gradually narrow one's choices down until the appropriate topic is offered and then accessed on tape or disc. More elegantly, a program can respond to "free-form" inputs which search for key words or patterns of letters. Questions like "What do you want to know about ——————" can be asked, and computer and/or video information presented according to what the user types in. New systems employing artificial intelligence techniques can interpret a variety of inputs, and even add to a program's vocabulary through interaction with users.

This interactive technique is best used as an adjunct to formal training—either as a reference for people who already know a bit about the subject—or as an aid for a "stand-up" lecturer. Instructors can effectively employ these visual data bases as illustrations for their classes, responding to questions or discussion points with specific examples. If a large number of visuals are needed, videodisc can be a compact way to store and access them. Capable of holding up to 54,000 individual frames per side, huge slide libraries can be contained on one simple vehicle.

Examples

Organizations have used videodisc to store such diverse images as medical slides, shots from NASA space missions, photography, and art from various galleries. The MITRE Corporation has explored the use of videodisc in storing medical records such as X-rays. Not only can great amounts of information be stored compactly, but the medium is easily and cheaply duplicated and is resistant to damage; therefore, individual users can each have their own copies of one program containing a wealth of visual information. Videotape can also be used to store visual data, but it is not as appropriate for single-frame storage and retrieval since not all interactive tape systems are frame-accurate, and holding one frame in "pause" tends to wear out the tape and clog video player heads. However, a number of short scenarios can effectively be stored on videotape and, combined with computer text, form quite a comprehensive instructional tool.

The United Way of Tompkins County, NY used an interactive video tape system driven by an Apple II computer to inform volunteers, community members, and representatives of member agencies, about its services and methods of operation. It was set up to be used in a public location, such as the United Way headquarters lobby and during the annual campaign at malls and businesses. The program displays a "teaser" of attractive visual scenes showing the various services of the United Way and connecting them to individuals. Then the computer posed the question, "What do YOU want to know about the United Way?" The program can "match" hundreds of word patterns and will branch the viewer to 15 main subject areas, and if the viewer wants more information, to several sub-segments of more specific information. If the user types in something that the program doesn't "understand," the computer requests that the viewer use a synonym or a more general term; if on the second try, the user's terms still don't match anything in the programmed-in lists, the computer displays a menu of segments from which to choose. The United Way has found that users spend quite a bit of time with this engaging system, and it seems to be more of an "information service" than the traditional "pitch." Figure 20 shows an example of the United Way flowchart.

Tompkins County United Way
INTERACTIVE INFORMATION SYSTEM

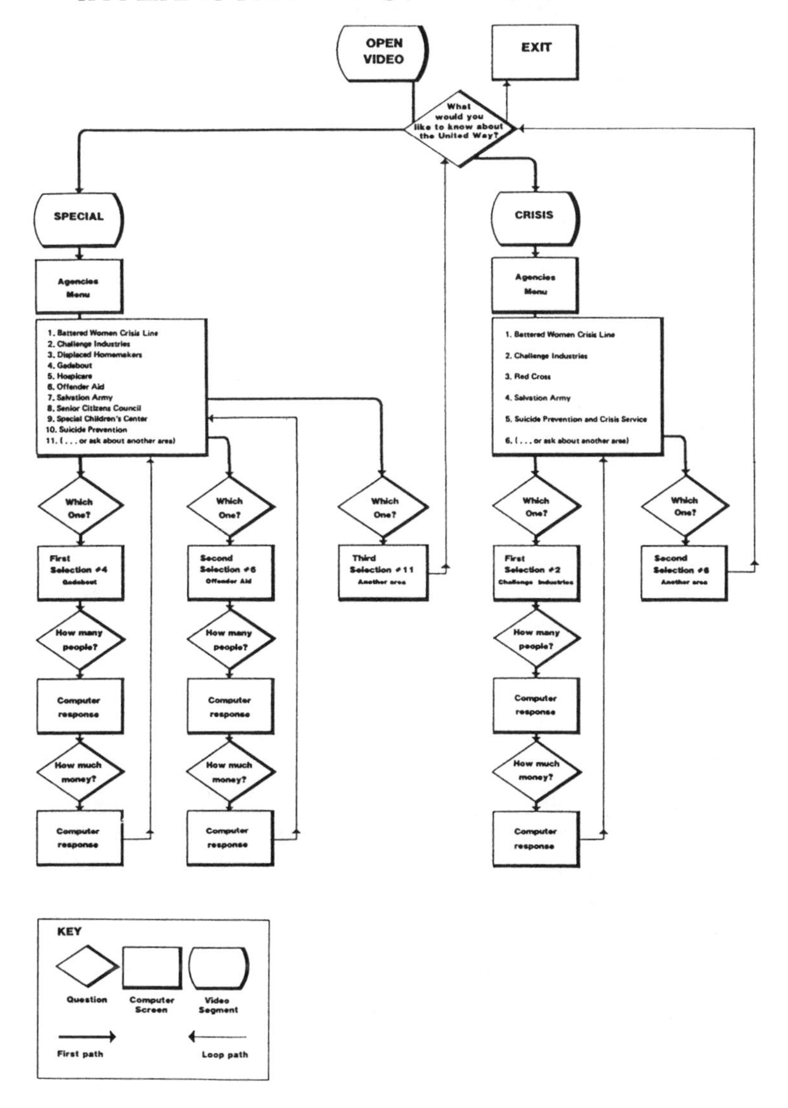

Figure 20. Example of flowchart for the United Way's visual data base.

PRETEST TECHNIQUE

Purpose

A pretest can be used to qualify trainees for instruction or to determine if they already have the prerequisite knowledge to make further information unnecessary. The pretest can be generated by computer text, and/or it can use visuals from videotape or videodisc. This technique is a time-saver in training, since the program can determine if the trainee has the prerequisite information necessary to make sense out of a videotape lesson; otherwise trainees can be branched to only the information they need—or test out of the program completely.

Description

Good pre-test design in interactive video involves primarily the same skills as paper-and-pencil test construction. Creating good test items is especially important here, since the results can either lock trainees out of further instruction or bypass further instruction completely. Therefore, it is necessary to have the test constructed and validated by someone proficient in testing and measurement. Of course, as in any training program, stating behavioral objectives first is a key to developing valid items. Where there is interaction among trainees who will take the pre-test over a period of time, it might be a good idea to develop alternate forms of items which can be randomly selected for each trainee (making sure, of course, that each alternate is similar in content and difficulty).

The testing can be done with or without feedback; that is, one can choose to inform trainees if they answered correctly, and if not, what the correct answer was. This can be done after each question, or after taking the entire test. In more sophisticated forms of testing, the test itself can "branch" depending on how the person is doing; this is called "tailored testing." The individual is started out at an intermediate level of difficulty, and gradually branched to easier material if he/she gets too many of the intermediate items wrong, or conversely, to more difficult items if a number of questions are answered correctly. In this way, trainees don't have to wade through obviously easy items, nor do they have to suffer through items at the end of a test about which they clearly have no clue. Time is saved and motivation (and energy) kept at a higher level when both testing and information-providing are presented only when necessary.

When developing tests, keep in mind the possibility and, perhaps, desirability of using still or moving visuals in items. Although this might seem obvious given the interactive video medium, most interactive programs still *teach* with video but *test* with computer generated print. Many trainees are simply not as adept at reading and writing interpretations as they are in correctly identifying and interpreting visual and aural cues. Instead of describing procedures or objects in wordy text, see if the same items can be shown more simply and realistically via video. Trainee's responses can take the form of short answers, multiple-choice selections, touching the screen, identifying a portion of the screen using a joystick or paddles, or in more elaborate systems, manipulating simulation devices like controls to equipment.

Pretesting is best used when the incoming knowledge base of the trainee population varies greatly, and there is prerequisite information needed to fruitfully learn from a given program. In some cases, an organization may want to ensure that trainees have mastered certain concepts, but don't want people to sit through material which they obviously know already. For new hires, this kind of training can be quite effective, since one can be confident that trainees have access to the information they need, but they can often go through the training process more efficiently since some whole sections or programs might be skipped over.

This technique is also extremely effective in teaching topics which people think they already know even if in reality, they *don't* (the know-it-all syndrome). If individuals think that they do know (or should know) some material, they often won't pay attention to a program. However, if by a pre-test it can be "proven" that they don't know it all, their motivation may be higher. They also may be cued about what information is really important, and pay better attention to relevant details in a subsequent program.

Example

Reliance Electric uses an interactive video program on basic aspects of gears for incoming sales engineers as well as for current employees. To heighten viewer motivation, a pre-test simulating a realistic but challenging customer interaction is used. Viewers try to solve the customer's problem and receive feedback about their choices, or skip the question and go right into the training. This pre-test not only challenges trainees but serves as an advance organizer for material to be presented in subsequent lessons. Reliance's videodisc on the basics of AC motors uses a game show theme with an introductory "qualifying round" to assess users' mastery of prerequisite information, followed by an "expert round" to weed out those who don't need the training at all.

SELECTING RELEVANT EXAMPLES TECHNIQUE

Purpose

This technique is used to make broad concepts more relevant and concrete to different individuals, based on their occupation, experience, interests, or personal characteristics. Through a menu or open-ended questions, individuals can decide which examples relate to them. This is similar to the "introductory menu" approach but the Selecting Relevant Examples Technique uses branching according to individual needs or roles of the trainee; the introductory menu technique branches according to concepts within the subject matter.

Description

"Selecting relevant examples" is an easy technique to use, and one that doesn't require elaborate or highly intelligent systems to implement. Users can branch to appropriate segments by scanning through a program until they see the appropriate boldly colored title, by finding the appropriate counter number on the playback machine or through a remote control device, by using a random-access device which searches out segments, or by a typical "interactive video" system using a built-in or external computer. If the program does not use an accurate interactive search device, be sure that a lot of room is left between segments, and that they are identified clearly by titles (making long title screens of distinctive colors can help when searching). Of course, this technique is only effective if there is a clearly defined audience, and one knows if there are, in fact, different examples that would be useful to present. Also, if one makes the effort to tailor the program, be sure to have appropriate branches for each "kind" of person who will watch it: for instance, if the program says "Would you like to see an example for (1) engineers, (2) sales representatives, (3) customers" and the person who is told to watch the program is an assembly technician, he or she will be offended or confused by the lack of an appropriate category. However, be sure not to stereotype classes of people so that they feel they are getting "second-class" or watered-down information. It's nice to give everyone the option of seeing every segment if they're interested–

just to let them know that there aren't any secrets that are only being shared with other groups. It's also often effective to let people see many different examples so that they can generalize concepts across situations.

This form of presentation tailoring is useful when there are different examples that people should see or might be interested in, and you don't want to take the time to show everybody everything.

Examples

Marine Midland Bank is using an interactive tape on "Professionalism" which includes tips on personal appearance. The program asks whether the viewers want to see the section on men's appearance or women's appearance: the men don't have to sit through minutes of demonstrations on applying make-up. In the women's section, a four-way split screen shows different women (one young blonde, a middle-aged brunette, a young black woman, and an older, gray haired woman) and allows the viewer to indicate which woman she'd like to see apply make-up. A woman can choose to see only the woman whom she preceives is most like herself, or opt to see several or all examples. This technique not only cuts down on training time, but it increases viewer's identification with the process. Marine Midland is also using this technique in an interactive program on bank security, since policies and equipment differ from region to region.

Carrier Corporation uses this technique with a simple remote-control videotape device that merely cues up a tape to a specified counter number and branches individuals to segments on a program on customer relations. The program has specific tips for service personnel, managers, parts clerks, sales persons, and secretaries. These individual examples help viewers learn specific techniques that they can immediately apply to their own jobs. Often it's difficult for trainees to translate broad ideas down to the level of daily performance. While they might understand general terms, they still might not practice what they've learned because the information was not specific enough—or they think that it doesn't apply to them. Following are portions of the Marine Midland interactive script and an excerpt from the Carrier Customer relations script.

Script Excerpt: Marine Midland

Segment 4

No job is pleasant unless people treat you well. Your job at Marine Midland is a professional one . . . . but do you sometimes wish your customers would recognize this? How *do* customers decide to treat a bank employee? Often, it's the way the banker looks and acts. That's what this program is all about—and we'll let you decide what parts you'd like to focus on.

(computer text)

The next portion of the program will show you how to achieve the professional look. But first would you like to see:

(1) men's appearance

(2) women's appearance

Script Excerpt: Carrier Corporation

We'd like to give you the opportunity to access a segment of this program that deals with your specific position.

The NARRATOR reaches off screen and pulls back a *responder* pad.

You will be able to do this by entering numbers into this auto search keyboard control. The numbers will appear on the screen next to a list of job titles. Simply pick the job title that you're interested in—and then enter its corresponding numbers into the keyboard attached to your video-tape machine. The keypad will then automatically access the segment of the tape that you chose. Are you ready to give it a try?

The screen cuts to a character generator page with four choices on it.

RESPONSE A SECRETARY/PARTS CLERK
RESPONSE B - BRANCH MANAGER
RESPONSE C - SUPERVISOR
RESPONSE D - SERVICE SALES ENGINEER

Next on the script will be the 4 response segments listed above, which the user may select individually or in sequence.

RESPONSE A - SECRETARY/PARTS CLERK

NARRATOR

It's important to make a good first impression to the customer—and usually that comes through answering the phone. Some Carrier Business Services branches employ secretaries as parts clerks also; in other branches the jobs are separated. No matter what your particular job description is, be sure that you get the customer started on the right foot.

Always identify your company when you answer the phone.

SECRETARY

Good morning, Carrier Business Services.

NARRATOR

If you don't, the customer may wonder if he got the right number.

SECRETARY

Hello . . . .

CUSTOMER

Is this Alaska Air Conditioning?

NARRATOR

Make sure that you identify the caller and the company he or she represents. Then try to . . .

VICARIOUS TRAVEL TECHNIQUE

Purpose

Interactive video can allow users to vicariously explore an area by making choices about what to see next, and in what detail. This technique is useful in allowing trainees to "pre-experience" a given location, saving time and expense in actual travel.

Description

Vicarious travel programs are usually menu-driven, allowing users to select what they want to see next by responding to a menu, or by manipulating a device like a joystick or steering wheel. Generally, an overview of a place is shown, and from there, options are presented to examine individual parts in greater detail. For instance, a general map of a town could be presented and the users can select which section they'd like to see first. Once the viewer gets down to an appropriate level of detail, the tape or disc can simulate actually being there, like walking down a street. As the person vicariously walks down the street, he or she may be given the option to stop the program at any point and examine an aspect more closely—like look at a building or even enter it. From there, the program can branch into even finer detail, exploring rooms in the building and even objects in the room. Obviously, such a program could contain an almost infinite level of detail, so that the producer must limit the domain of experiences presented. Often, these programs are presented on videodisc, or even "ganged" videodisc players so that a sufficient number of scenes can be stored and so that branching among these scenes can take place rapidly. This is important because the "being there" depends on realism in terms of the time it takes to go from point A to point B. In exploring towns and buildings, for instance, shots would have to be taken walking down each street or hall, and segments made of each choice possible—turning right, turning left, or even turning back. These productions are often done with special video equipment with four cameras mounted on a tripod, shooting the "north" "south," "east" and "west" views of each scene. Needless to say, this is a lot of footage to be shot. Usually these programs have little or no "training" dialogue; they are designed to simulate as closely as possible the actual experience of being there.

Examples

Probably the most famous vicarious travel program is MIT's Aspen disc, which explores Aspen, Colorado; this has provided the model for a number of working programs in industry and the military. For instance, C-E Power Systems is producing photo-documentation videodiscs on various nuclear power plants using their CEVUE system. This technique is used to assist nuclear plant personnel in learning their way around a facility. Emergency conditions can easily be explored, such as certain routes being blocked off, and trainees can practice how they'd deal with those contingencies. The applications for military intelligence and tactical training are quite obvious. This approach has also been used to allow users to explore museums and examine details of exhibits, making a fascinating self-directed study piece. Figures 21 and 22 on the following two pages show a diagram of the CEVUE system.

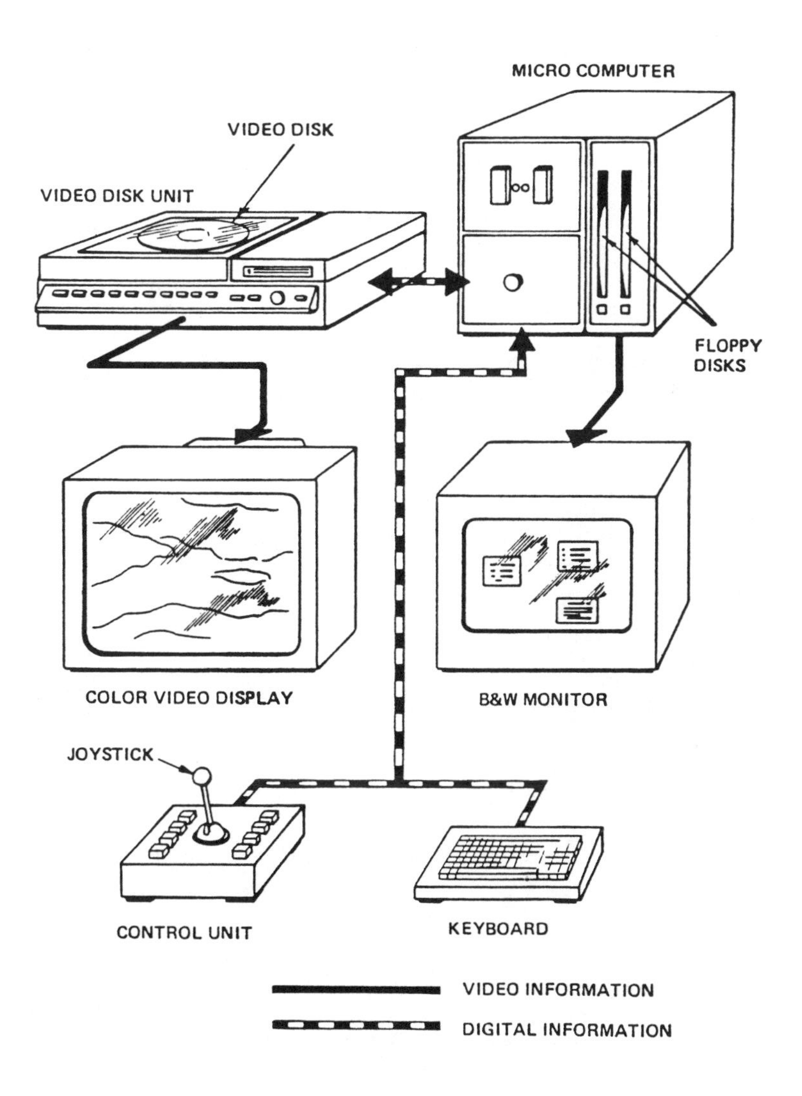

Figure 21. Hardware connections for CEVUE vicarious travel program, courtesy C-E Power Systems, Inc.

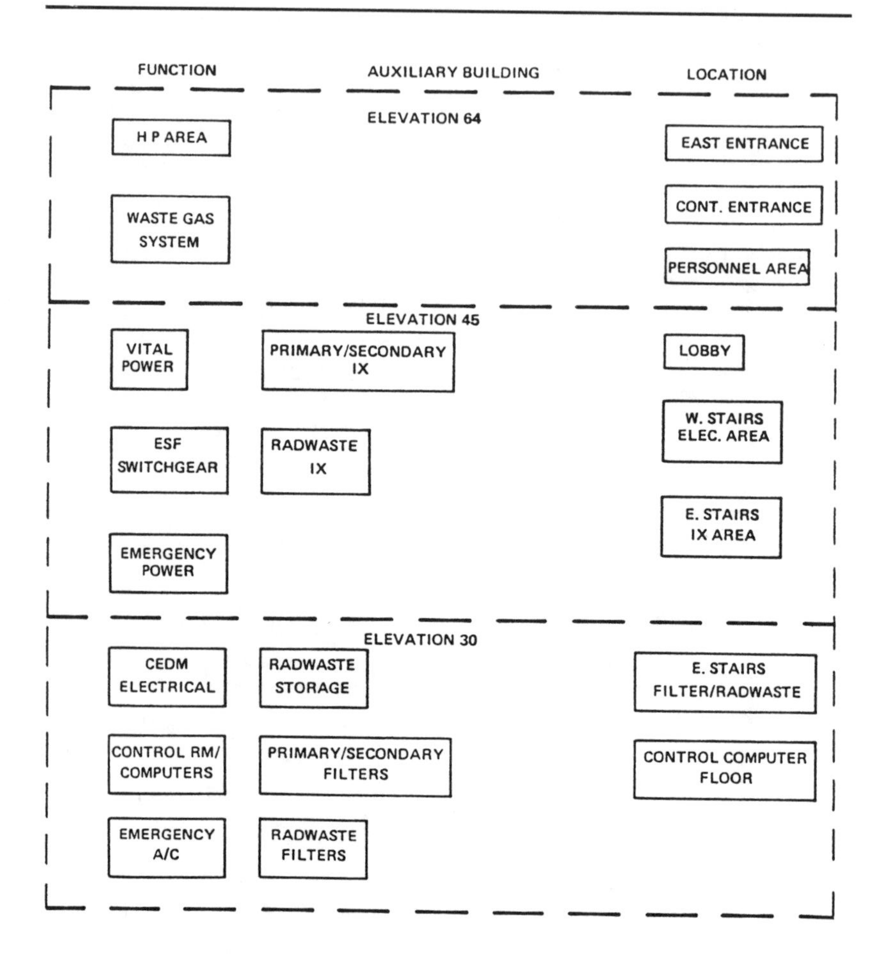

Figure 22. Diagram of locations within C-E Power Systems photodocumentation videodisc program, courtesy C-E Power Systems, Inc.

THIRD PERSON "DIRECTED" SIMULATION TECHNIQUE

Purpose

A "directed" simulation is used to allow trainees to experiment with concepts and procedures by letting them vicariously take part in a simulation. Simulations are useful in allowing trainees to practice the information they've learned, and in testing their ability to apply that information.

Description

A "third person" simulation is one in which the trainee "directs" someone on the screen what to do, and then sees what happens. A program is designed to set up a hypothetical situation among a number of characters; at various places in the situation, choices emerge. Through a keypad or keyboard, the user decides what should happen next, and sees the consequences. The trainee may not be given any specific feedback about whether the choice was right or wrong at each step; in a true simulation, no all-knowing narrator comments on the trainee's decision. It's up to the trainee to decide for him or herself if the choice was a good one as the scene is played out. This technique is similar to "behavior modeling" discussed earlier in the linear video section; here, the trainee has a say in the model behavior and outcomes.

In good simulations, there must be extensive research carried out first to see exactly what choices are often made in similar situations, and what the consequences are of such choices. It's important for the simulation to be believable and justifiable on the basis of facts. Simulations must have a number of decision points, and be complicated enough to be intriguing: one wrong reply shouldn't spell necessary disaster—even in real life, most of the time we can manage to recover from mistakes if we catch them fast enough! The options presented must be attractive enough so that the user doesn't always pick out the correct path: otherwise, most of your branches will never be seen.

The "third person" approach is good when you want trainees to explore both the "right" and the "wrong" ways to do things so that their general concepts are strengthened. Often viewers will enjoy "making" someone in the scene do something "stupid" when, in a more direct simulation, they may not want to "themselves" make the same mistake. This exploration can not only be fun, but can increase the time and attention devoted to a particular program.

Examples

These simulations are often used in training for situations which are high stress or dangerous—either financially, physically, legally, or emotionally. Interpersonal skills like selling or customer relations techniques are often engagingly presented in this form. Viewers can "see what they look like" by watching another person in their role behave in different ways; this more objective viewpoint can aid people in seeing themselves more realistically and perhaps in breaking bad habits. For instance, a commercial provider of training materials for banks produced a third person simulation on bank selling skills. After being introduced to basic techniques, viewers watch the opening of a typical business call which introduces the possibility to sell some bank service. The tape stops, and a question screen asks, "What should Bonnie do next?" and presents several alternatives. The trainee then sees the results of this or her choices carried out.

Ford used a third-person simulation of golfing on various famous courses throughout the world to introduce its videodisc network to dealers in a lighthearted manner. Users get to choose which club they'd like to use, and see the effect of its use at each hole. The National Library of Medicine has produced a videodisc on teenage suicide containing third person simulations of patient interviews.

FIRST PERSON SIMULATION TECHNIQUE

Purpose

The purpose of the First Person Simulation Technique is to recreate as closely as possible an actual situation with which trainees will have to deal. Like the other forms of simulation, it is most effectively used after the learner already has mastered some basic concepts, so that he or she doesn't just flounder around guessing at probable strategies. This realistic simulation can be used as a substitute for traditional role-playing to reinforce skills and to test for the acquisition of basic skills and concepts.

Description

In a first person simulation, the trainee "directly" interacts with a person or a group on the screen. The scene is shot in such a manner as to simulate how a real scene would appear through the eyes of the user if he or she were actually there. By responding to questions or menu options, the learner decides what to do at each of several (or many) decision points. No outside feedback is given as to whether each choice was correct or incorrect; the trainee must learn to judge this for him or herself. Therefore, this kind of simulation not only practices skills, but aids in building the ability to evaluate oneself.

Again, research into "real life" examples must be done first so that options presented are attractive and believable, and to determine that the consequences presented do, in fact, occur. First person simulations are very involving for the viewer—but only if they're believable from their realm of experience. The situations must be complex enough to involve a number of decision points, and are most interesting if they are not merely "good" or "bad" choices. In real life, things are rarely so black-and-white, so trainees should have the chance to recover from a mistake, or to merely do an "acceptable" job.

This approach is used as a substitute for, or adjunct to, role-playing. In traditional role-playing, trainees are usually paired off, for instance, one playing a customer and the other playing the sales rep. While in sales training, they might have gotten a lot of information about how to act as a sales rep, usually no one is taught how to "play" a customer accurately; therefore, many of these role-plays among students are less effective than they could be. Furthermore, when students are paired off like this, often they will play the roles too "easy" or too "hard" because of the nature of the training situation and their acquaintance with their partner. Videotaped "role-play partners" can not only bring more authenticity and variety into the roles, but also can be used by individual trainees. The trainee can play the "salesperson," or the tables can be switched to let the trainee play the customer to learn what it feels like to be in the other person's shoes.

In order to produce such an interactive program, the videotaped respondant must be shot reacting to each of the questions or comments offered in the options. Often, these scenes will have *multiple dependencies* which means that if one option is selected, it may very well eliminate others. In other cases, *when* a question or comment is raised makes a difference in the consequences. For instance, in sales techniques, it's certainly a good idea to try to close the sale—but that must neither happen too early nor too late in the exchange.

Examples

This approach is used effectively in sales and customer service training, as well as in teaching various interviewing techniques. For instance, the Career Planning Center at Ithaca College uses a simulation of a job interview; here, the tables are switched, and the student/viewer gets to act as the job interviewer and ask a potential candidate some questions. By doing so, they can better put themselves in the other person's shoes and understand the kind of information a recruiter looks for. The University of Helsinki Psychiatric Clinic uses a first-person simulation to teach medical students about psychiatric diagnosis: If the student takes the wrong approach, the patient will either deny his problem or clam up completely; if the correct method is chosen, it's discovered that the patient has attempted suicide. Another engaging program produced by an interactive video class at Ithaca College simulates a situation in which the viewer as sales clerk in a department store has to handle a customer who wants to return a defective food processor. If the clerk is too rude, the person leaves in a huff; if the clerk is too lenient and gives the customer his money back without a receipt, we find out that the whole thing was a scam and see the customer and his cohort planning their next "job" in the parking lot. If the correct decisions are made, you politely refuse to return the money until the receipt is presented and watch the customer in the parking lot angrily telling his buddy that they should have tried another store.

SIMULATION WITH FEEDBACK TECHNIQUE

Purpose

In this technique, trainees can experience a situation while getting explicit feedback about their choices along the way. While "true" simulations might be confusing for novices because they might not be able to evaluate their own decisions, this approach gives them more immediate information about their choices and why they were right or wrong.

Description

This kind of simulation can either be first-person or third-person (the trainee can either interact directly with a person on the screen or "direct" someone on the screen how to act). A situation is broken down into a number of decision-points; after the user makes a decision, the consequences of that decision are shown and a narrator comments on what's just happened. Therefore, in production, one must not only produce a number of possible scenarios, but also create some explicit feedback and analysis. As in the other kinds of simulations, the consequences must be based on real data and must be believable to the audience. The feedback might include letting the viewer go back and try that choice over, or automatically showing him or her the right way to do it.

Simulation with commentary is an easier way to introduce trainees to a new subject since they always know the reasons why their decision was good or bad; in some complex situations, without commentary, they might be confused or come to the wrong conclusions about why the situation turned out as it did. While true simulations are more suited to practice and testing than to instruction, simulation with feedback can be used to introduce trainees to information which might be partially unfamiliar. Through the simulation, like a pre-test, trainees might become more aware of their own shortcomings in knowledge and skills, and be more motivated to learn the material.

Examples

Several demonstrations of interactive video technology have made use of the simulation with feedback approach. The demonstration program for the Panasonic Interactive Video System on "Office Skills" contains a sequence on how to handle visitors with and without appointments in the office. A situation is described via a text screen and the viewer is asked how it should be handled. Depending upon the response, the viewer is shown what would happen if that choice were made; after a few seconds of the dramatized scene, a voice-over narrator describes why the choice was correct or incorrect. Another demonstration program produced by BCD Associates, manufacturers of interactive video interfaces, simulates medical intervention for witnessed and unwitnessed heart attacks. If the students pick the correct treatment at each step, a physician on camera follows their direction until the next decision point is reached. If the wrong choice is taken, a narrator explains what should have been done and the previous sequence is played over so that another choice can be made.

GAMING TECHNIQUE

Purpose

Gaming is a variant of simulation approaches in which a trainee competes against the interactive system, or groups of trainees compete with each other using the video system as a basis for information and scoring. It is used to heighten trainees' attention and motivation by arousing their natural competitive instincts and to add a novel dimension to training.

Description

The gaming approach can be used with almost any other interactive technique, and is a matter of how questions and evaluation are treated. There are several basic strategies that can be employed: (1) Having the computer ask questions of teams or individuals, keeping their names and scores in a scoreboard fashion. This kind of program would start out asking for individual or team names. As questions are asked, they can be directed to one of those names, and scores kept according to those pre-defined categories. This sets up a game between individuals or groups participating together; (2) Using some aspect of randomization, either in choosing questions, turns, or situations. This technique can use the ability of a computer to generate random numbers to pick a questions, or an outcome. For instance, the computer might randomly decide whether or not a customer would respond well to a trainee's approach (just as customers tend to do in real life), or the program could randomly choose which of several scenarios to present. This can avoid having each trainee in a group going through exactly the same program, and enables an individual trainee to have several different experiences with the same program; (3) Presenting a problem which an individual and the "computer" compete to solve. This is similar to number 1, but in the absence of other trainees, the computer can "play" the opponent and compete with the trainee. For instance, the computer might randomly or based on some pre-defined pattern choose a response to a situation; the trainee can also choose, and then presenting the outcome the trainee can see how he or she fares compared with the computer. This is a difficult technique to master, except in situations such as playing strategy games similar to chess; and (4) Presenting response evaluation and feedback in a traditionally game or sports-like fashion using light-hearted video and/or audio. This involves no special computer programming, but does involve a different flavor for video feedback, relying on typical sports scenarios with which the trainee population might identify. For instance, a correct answer might be met with a stadium crowd cheering, or typical sports victory music. Small groups can use one keyboard or individual response pads, or a trainer can key in individual's responses for them. Segments showing scores on a scoreboard, sports music, or video game-like graphics can be presented depending upon the answer.

Gaming is best used for small groups who are relatively well acquainted with the subject matter and with each other. It's most effective in situations where competition is expected—such as sales—but would be quite tasteless in such areas as emergency medicine. Like other games, it's only fun if all participants have a fairly equal shot at winning. Unless designed carefully, though, it can diminish the seriousness of what's being taught, and is probably good as an ice-breaker or instructional diversion amidst more traditional training approaches.

Examples

Learning International (formerly Xerox Learning Systems, Inc.) offers a version of its sales course, Professional Selling Skills System III incorporating the use of a videodisc. The program is instructor-led and uses the videodisc to present still and motion examples. Twice during the course, gaming sequences are presented during which teams of trainees compete to answer questions. The trainer keys in a code depending upon whether the response was correct or incorrect and the disc presents the score and some lighthearted feedback. Following is a photo from the program.

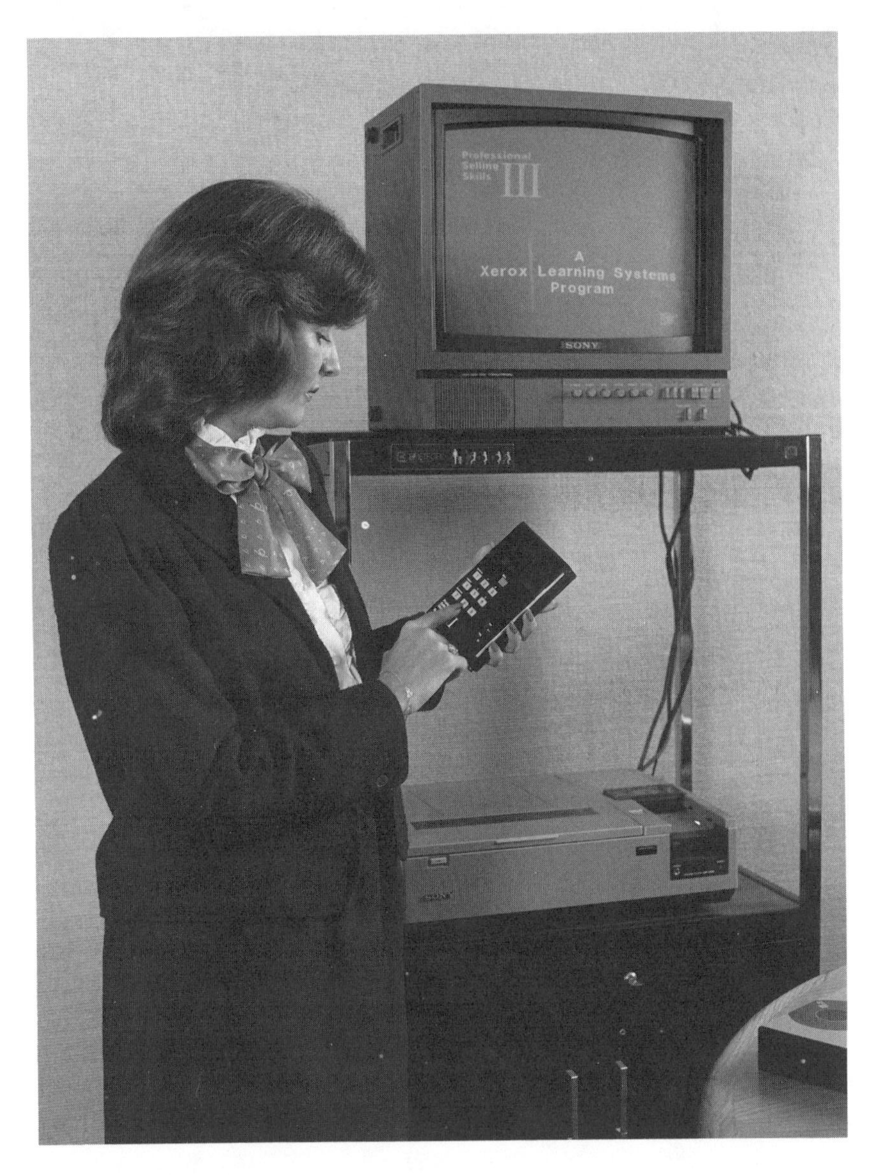

Figure 23. Trainer using videodisc controller with the Professional Selling Skills videodisc-assisted training program, courtesy Learning International, Inc.

TUTORIAL TECHNIQUE

Purpose

The tutorial is the paradigmatic form of interactive video, where the program engages in a dialogue of information and questions with the student. A tutorial is the most common, and often the most appropriate way to introduce complicated new material since it takes the subject step-by-step, ensuring that the learner understands the previous point before going on. This format attempts to recreate the "Socratic" teaching method, teaching by questioning and actively leading the learner to uncover the basic concepts and information. Of course, an interactive tutorial gives immediate and specific feedback to each question, and can record learner responses for later analysis by a trainer.

Description

In a tutorial, the material is broken down into concepts, usually by first doing a task analysis and writing specific behavioral objectives. Then the designer must find out not only what the "correct" content is, but also what mistakes are common to people learning these concepts. These different mistakes or response patterns form the bases for questions and feedback loops. If a trainee exhibits some difficulty in learning a certain point, he or she may be branched to some smaller, easier, and more basic segments, or an explanation of the content in different terms. Learners who answer correctly can be presented with short, more sophisticated explanations, so their viewing time will be shorter than those who have difficulty with the topic. Therefore, various kinds of explanations for each topic must be produced, some taking it quite slowly, others at a moderate pace, and still others using the briefest language possible. Trainees who show themselves to be able to "catch on" quickly can be branched to the most concise explanations.

In tutorial approaches, it's important to not merely tell the trainee that he or she is "right" or "wrong," but also to explain *why*. Rather than just "zap" them and make them watch the same segment over and over until they get it (or more likely, guess it), they should be branched to specific segments that address their misconception and acknowledge why they might have been led to answer as they did. Once understanding the basis for their mistake, they can be asked to try again, or merely given the correct answer.

Examples

The Tutorial Technique is good for teaching complex concepts, like mathematical or scientific ones, or for assisting learners to memorize a great deal of information, like details about a company's product line. For instance, Reliance Electric uses the tutorial approach in their interactive video programs which teach new sales engineers about the working of their large line of electromechanical products. In a program on maintenance of bearings, for instance, a section explains the necessity for preventive maintenance and the tools and concepts needed to support it. One particularly difficult topic is how to read the charts to decide the proper temperature rating of grease to apply in different situations. After a general introduction, a computer graphic shows a chart and asks the trainee to decide on the temperature rating for a particular case. The computer can analyze ranges of numerical inputs and explain why choices were correct or incorrect. Figure 24 curtains a portion of the Reliance "bearings" flowchart.

The Tompkins County (NY) Fire Department has produced an interactive video tutorial on the new radio communications equipment. Since most firefighters are volunteers, it was necessary to provide through instruction at a wide variety of times and places. Using a simple random access controller, the users answer questions about each aspect of the radio's operation as the topic is presented step-by-step.

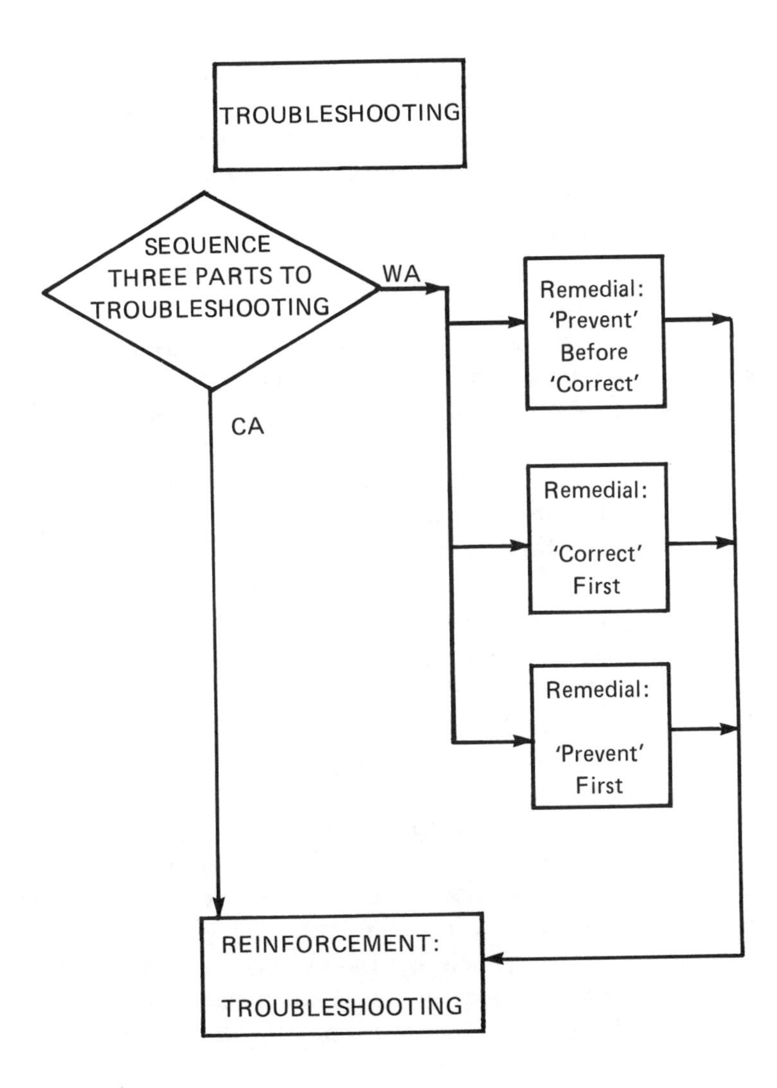

Figure 24. Reliance Electric Co. flowchart showing 4 branches to a question on troubleshooting, 3 for wrong answers showing remedial segments, and one for the correct answer providing reinforcement. Courtesy OmniCon Associates.

OPINION QUESTION TECHNIQUE

Purpose

Questions can be useful for uncovering individuals' opinions and values on a topic and relating the content in light of those feelings. It is used to heighten motivation, make the information more personal, and simulate more closely human communication where the affective domain is as important, or more important than, the cognitive domain.

Description

Often interactive video designers fall into the trap of designing questions which only assess learners' mastery of content. Although this recreates one part of interpersonal communication, it misses an important part of training: assessing individuals' opinions and values on a topic. Once a designer understands "where the trainee is coming from," he or she has a better chance at directing the message successfully. Training programs often are created to change attitudes—so it's important to first assess what attitudes are held. For instance, if one were designing a program on compliance with safety measures, it would be a good idea to find out what individuals think of current safety regulations, and if they really do practice what they know.

If the writer can find out what is important to a trainee, the program can point out how the instruction applies to his or her own interests. This is the kind of introductory "patter" that often characterizes good stand-up training, and is more important than a throw-away warm-up exercise. Determine what kinds of attitudes and motivations may be important in presenting your content, and ask some introductory questions to assess them. Then, depending upon the responses, present branches which acknowledge those feelings and put them into the perspective of the training. For instance, on a program on phone sales, the program could ask how the individual really feels about the issue. Training a person who thinks phone sales is a bad idea is much different from presenting that material to someone who doesn't know much about the topic or to someone who is already using phone sales techniques successfully.

The privacy of the interactive medium can allow the presentation of questions which would be too embarrassing to ask in a group of one-on-one situation. Be careful, though, in recording responses: if individuals are to be honest in expressing their feelings, be sure to let them know that their responses are NOT being recorded, or if they are, that the printout is for their use only. People will be much more hesitant in letting you know their real opinions if they think it may wind up in their files! Of course, principles of good questionnaire design apply here. Be sure not to "load" the questions with only one response that is socially acceptable.

Examples

Consolidated Diesel produced an interactive video program to introduce this new instructional technology to trainers and managers. As a part of the presentation, users were asked if they had ever seen interactive video before, and what they thought about using it as a communication tool in their own units or content areas. Responses were anonymous, but collected on diskettes for review by the media staff and interactive developers.

An interactive video program on professionalism produced for Marine Midland Bank starts with a question asking what goals are most important to the user's career. This information is used to branch the viewer to a segment relating the program's objectives to the individual's interests.

AUDITORY/VISUAL TEST TECHNIQUE

Purpose

Interactive video can not only present content in an audio-visual format, but it can test in this manner, too. The use of images and sounds to present questions provides a new range of content that can be covered, and enables the trainer to evaluate responses in a manner similar to the way in which the content was taught and in which it will be applied.

Description

Too often training is presented using the rich stimuli of color, moving visuals, the human voice, and sound effects, but the evaluation of learning is based on reading and writing cumbersome text. In many training applications, trainees may not be fluent in reading or writing, and therefore may score poorly on items which they actually know. Many topics are also much more easily tested using audio-visual examples.

When using interactive video, don't assume that all questions will appear in a text form: examine the materials and see how items might be presented in video. For instance, a program might show three short clips of a salesperson handling an objection, and then ask which example was correct. One might present four tools and ask which is the correct one for the job. Using audio, the sound of an engine with a bad valve can be presented, and the program can then ask the trainee to identify the probable cause of the problem. So that the choices remain visible for the trainee to consider, hold a still shot of the choices (perhaps using a split screen), freeze-frame the video during post-production, or use an interactive system which automatically freezes the tape or disc in playback.

One difficulty with auditory-visual tests is that often the question can't be repeated so that the trainee doesn't have the same opportunity for examining the question as he or she might with print or computer text. A way to get around this is to offer a way to repeat the question; a small cue line can appear at the bottom of the question screen. For instance, the program might include a split screen of three different salespersons freeze-framed after each had just handled a customer's objection. Although each salesperson might look distinctive enough, the trainee might forget which one said what. So, at the bottom of the freeze-frame, key in a line saying "to repeat question, press 4." Pressing 4 would repeat each vignette and give another opportunity to answer the question with no penalty.

An interesting application for this technique is in standardized testing which seeks to measure individuals' sensitivity to non-verbal cues or one's ability to comprehend motion and relative positions. For instance, one might construct a test of mechanical ability which presents moving objects such as gears and asks about their relationships and about how changes would affect them. Another example would be in measuring trainees' ability to read "body language" and tone of voice. This kind of test can be used either after specific instruction, or as a screening device.

Examples

More traditional uses of visual testing are found in industrial training for assembly and quality-control. For instance, NCR's Ithaca, NY plant used an interactive program to train line workers to assemble and inspect printers. One question showed a few parts alongside one another and asked which one should *not* pass inspection. The program taught individuals how to distinguish between insignificant variations and important deficiencies in parts. ICS-Intext uses visual tests in its training program on multimeters produced by Perceptronics; one question, for example, shows four types of meters in four quadrants of the screen and asks the trainee to touch the VTM (one particular type of meter). Using a touch-sensitive screen, trainees can pick out a meter and receive feedback about the correctness of their choice. Figure 25 shows an excerpt from the ICS lesson design form showing this question.

3.1.5-2 MULTIPLE CHOICE WITH TOUCH ActionCode™ IDS

SEQUENCE: 2 TOPIC: 2 LESSON: B0102,3 LEVEL: 1 DESIGN: 5/28/83
TITLE: Identifying Multimeters APPROVAL: / / 83

VIDEO SEGMENT V1: | REFERENCES
TYPE: [ ] Motion [ ] Slow Motion
[ ] Stepped Stills [X] Single Still
LENGTH: [ ] Short [ ] Med [ ] Long
Stills: 1
CONTENT (Describe or Insert):

VOM, VTVM, TVOM and FET VOM, each in a different quadrant.

Overlay: Point to the VTVM.

Make sure distinguishing features are clear.

AUDIO [ ] 1 or [ ] 2 (Describe or Insert):

TOUCH SCREEN T1:
PATTERN:
Pictures: [ ] 2 [ ] 3 [X] 4
Text: [ ] 5 [ ] 6 [ ] 7 [ ] 8
Diagram: [ ] 12 [ ] 13 [ ] 14 [ ] 15
FORMAT: (Describe below):
CONTENT (Describe or Insert):

POINT TO THE VTVM

METER A	METER B
METER C	METER D

Figure 25. Excerpt from lesson design form for visual testing sequence. Courtesy of Perceptronics, Inc.

RESPONSE PERIPHERAL TECHNIQUE

Purpose

Response peripherals are used to allow trainees to respond in more natural ways than by pressing buttons or typing on a keyboard. By using external devices, individuals can answer questions or indicate menu choices by pointing, touching, or moving some object and thereby more closely simulate on-the-job conditions. These peripherals can also make it easier for those who find it difficult to type, or whose hands aren't free to use traditional response mechanisms.

Description

Response peripherals are categorized together as the fifth level of interactivity and range from relatively common touch-screens, joysticks, and light pens through bar-code readers to full-blown simulators. They can be used as alternative input mechanisms for responding to questions, as part of any of the other interactive techniques listed, and can additionally provide a means to evaluate psychomotor responses. For instance, one paper mill uses voice-recognition to register trainee responses because certain employees have difficulty in writing and typing. A series of mass-marketed training discs uses a bar-code reader as a computer peripheral to tie workbook questions and examples to videodisc segments: the trainee merely scans the appropriate bar code printed underneath a choice in the workbook with the attached bar-code reader (about the size of a pen) and the video branches to an appropriate scene. Here, the bar code is read and analyzed much like a trainee's typed-in response might be. More sophisticated systems use simulator-like devices to teach flying, tank gunnery, and CPR.

Developing response peripheral systems can become quite complex, especially in the cases where special devices must be engineered and built. Touch screens, joysticks, paddles, and light pens use standard x-y coordinates that identify points on the screen to register a response and evaluate it. These points are common referents even in such widespread languages as BASIC. Remember, though, that this evaluation is basically in a multiple-choice form. Although these systems are not much of a problem from a design or programming point of view, they do add to the cost of the delivery hardware and are one more item that needs maintenance. Systems like those used for flight training or CPR involve electronic engineering and fabrication of sophisticated manipulanda that not only mimic real-life objects, but also stand up to the rigors of training.

Custom-designed or even off-the-shelf response peripherals are worth the time, cost, and effort if a significant training need exists that falls mainly in the psychomotor domain. They are inexpensive in comparison to providing real planes, real tanks, or real heart-attack victims, yet afford trainees the opportunity to "pre-experience" a situation. These computer systems are often more accurate in measuring trainee behavior than are live instructors, and with the use of videodisc, can provide almost instantaneous feedback to responses.

Examples

Perhaps the best-known interactive system and certainly a classic example of using a response peripheral is the CPR system developed by the American Heart Association and now marketed by Actronics, Inc. The system combines a specially-wired manikin, an Apple computer, two monitors, a disc player, and a random-access audio deck. Using it, people can learn CPR in a matter of hours or days, and get specific and accurate feedback about the accuracy of their compressions and resucitation. The system includes an exam for AHA-approved certification and was found to be superior to traditional class instruction. An additional module, "Megacode" teaches advanced life support techniques for emergency room personnel. Trainees respond to questions and simulations using a light pen to touch points on the screen, or actually apply compressions and provide respiration to the manikin. Sensors in the manikin can "tell" whether trainee's compressions and respirations are correct, and send that information back to the computer which in turn, causes the videodisc to branch to appropriate feedback. Following are a photo and a portion of the script (Figure 27) from the CPR system.

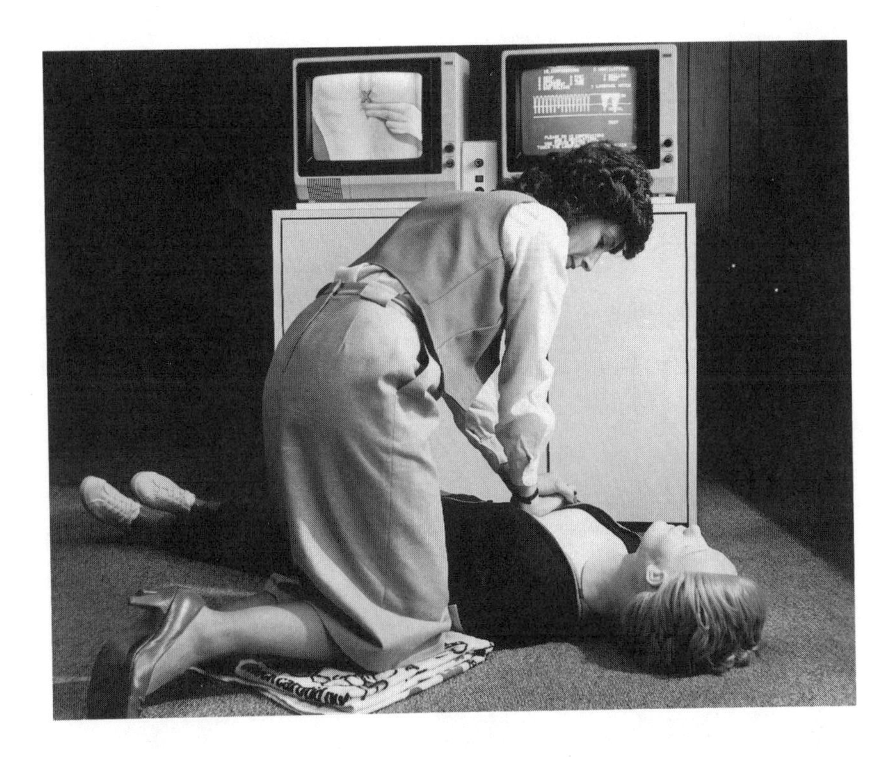

Figure 26. Photograph of trainee using CPR response peripherel videodisc system. Courtesy, Actronics, Inc.

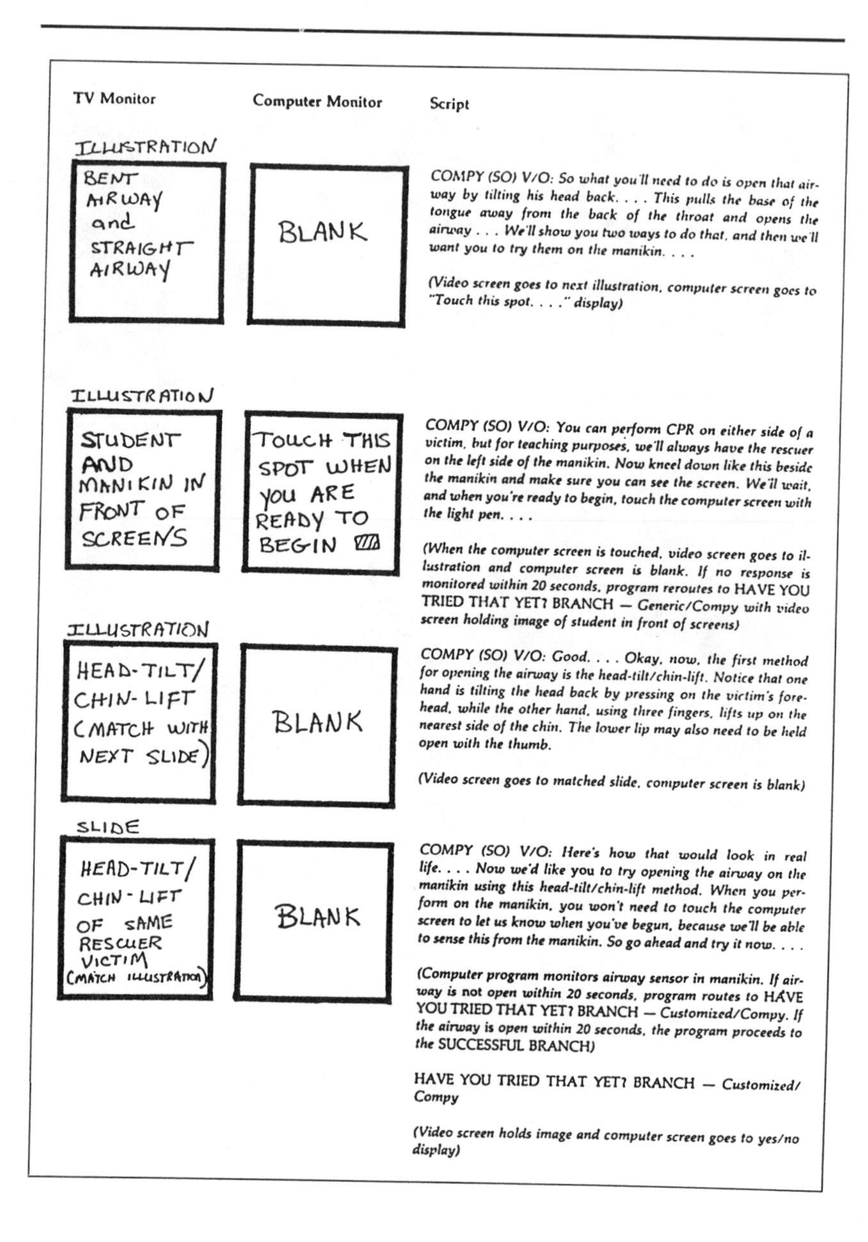

TV Monitor	Computer Monitor	Script
ILLUSTRATION: BENT AIRWAY and STRAIGHT AIRWAY	BLANK	*COMPY (SO) V/O: So what you'll need to do is open that airway by tilting his head back. . . . This pulls the base of the tongue away from the back of the throat and opens the airway . . . We'll show you two ways to do that, and then we'll want you to try them on the manikin. . . .* *(Video screen goes to next illustration, computer screen goes to "Touch this spot. . . ." display)*
ILLUSTRATION: STUDENT AND MANIKIN IN FRONT OF SCREENS	TOUCH THIS SPOT WHEN YOU ARE READY TO BEGIN	*COMPY (SO) V/O: You can perform CPR on either side of a victim, but for teaching purposes, we'll always have the rescuer on the left side of the manikin. Now kneel down like this beside the manikin and make sure you can see the screen. We'll wait, and when you're ready to begin, touch the computer screen with the light pen. . . .* *(When the computer screen is touched, video screen goes to illustration and computer screen is blank. If no response is monitored within 20 seconds, program reroutes to* HAVE YOU TRIED THAT YET? BRANCH *— Generic/Compy with video screen holding image of student in front of screens)*
ILLUSTRATION: HEAD-TILT/ CHIN-LIFT (MATCH WITH NEXT SLIDE)	BLANK	*COMPY (SO) V/O: Good. . . . Okay, now, the first method for opening the airway is the head-tilt/chin-lift. Notice that one hand is tilting the head back by pressing on the victim's forehead, while the other hand, using three fingers, lifts up on the nearest side of the chin. The lower lip may also need to be held open with the thumb.* *(Video screen goes to matched slide, computer screen is blank)*
SLIDE: HEAD-TILT/ CHIN-LIFT OF SAME RESCUER VICTIM (MATCH ILLUSTRATION)	BLANK	*COMPY (SO) V/O: Here's how that would look in real life. . . . Now we'd like you to try opening the airway on the manikin using this head-tilt/chin-lift method. When you perform on the manikin, you won't need to touch the computer screen to let us know when you've begun, because we'll be able to sense this from the manikin. So go ahead and try it now. . . .* *(Computer program monitors airway sensor in manikin. If airway is not open within 20 seconds, program routes to* HAVE YOU TRIED THAT YET? BRANCH *— Customized/Compy. If the airway is open within 20 seconds, the program proceeds to the* SUCCESSFUL BRANCH*)* HAVE YOU TRIED THAT YET? BRANCH *— Customized/ Compy* *(Video screen holds image and computer screen goes to yes/no display)*

Figure 27. Excerpt from script where response peripherals are used to teach CPR. Courtesy Actronics, Inc.

TV Monitor	Computer Monitor	Script
Continued	☐ Yes ☐ NO	*COMPY (SO) V/O: Did you try it? If you did, answer yes on my screen. If you didn't, answer no.* *(Depending upon response, program advances to* NO BRANCH *or* YES BRANCH*)* (NO BRANCH) *(Video screen holds image of head-tilt/chin-lift, computer screen goes blank)*
Continued	BLANK	*COMPY (SO) V/O: Well, don't be bashful. Please go ahead and try opening the airway. This unconscious victim needs air to breath.* *(Program continues to return to this response if no action is indicated)* (YES BRANCH) *(Video screen holds image of head-tilt/chin-lift, computer screen goes blank)*
Continued	BLANK	*COMPY (SO) V/O: Well maybe you need to tilt the head back more, and lift up on the chin a little more forcefully. Please don't be hesitant—we want the airway wide open so that this person can breathe. . . . Once you've got it right, we'll proceed—Now, go ahead and try it again.* *Program monitors performance and proceeds to appropriate branch. If two "Yes" answers entered in a row, program reroutes to* LACK OF SUCCESS BRANCH—*Generic/Compy. When airway opens, program proceeds to* SUCCESSFUL BRANCH*)*
ILLUSTRATION In bold, highly elaborate script: "Excellent!"	BLANK	(SUCCESSFUL BRANCH) *(Video screen holds image, but adds TV key, computer screen goes blank)*
(SAME AS EARLIER head-tilt/chin-lift) WE USE THIS TECHNIQUE IF THE VICTIM IS WEARING DENTURES	BLANK	*COMPY (SO) V/O: Excellent . . . And if the victim has dentures, this is the method to use to hold them into place. If rescue breathing is needed, the airtight seal is easier to perform when dentures are in place. But, if dentures cannot be managed in place, remove them.* *(Video screen goes to illustration, computer screen is blank)* Listing 1 shows the programming that brought much of this excerpt to the video screens.

Figure 27 (continued).

USER COMMENT FILE TECHNIQUE

Purpose

With most interactive systems, trainees can only "speak" when they're spoken to. Using a "comment" feature with systems incorporating microcomputers, you can allow users to type back comments to you *ad lib*. These comments can provide useful feedback about the software and hardware so that the training system can be fine-tuned and so that you can learn more about the individuals participating in the training.

Description

To provide a means for trainees to leave comments to you, the computer program must be structured to regularly "look for" a certain key to be pressed. This is usually a specially designated function key or pressing a control character with another key. When the program detects that key press, it branches to a subroutine that stops the program, clears the screen, and allows the user to type in a free-form message. Upon another key press, the program resumes.

This feature is especially helpful when field-testing programs and it's not certain if all the "bugs" have been worked out—either technically or instructionally. It essentially provides the same function as raising one's hand in class. Learners can leave a comment about a problem in the lesson execution, or something they felt was especially effective or ineffective, or they can request further information or clarification. When the trainer, developer, or course manager goes through the class records on diskette or paper, he or she can note those responses and take appropriate action.

Of course, this is an unusual option, and trainees will have to be told how and when to use it. It's appropriate to give them some practice using it in the beginning of a program so they know how it works; they will then be more likely to use it when a comment crosses their minds. This feature can be built into any interactive system that is custom-programmed in a programming language, and is part of at least one authoring system.

Examples

This technique can be applied to almost any training situation in which the hardware is available and the authoring aid supports the feature. It is recommended as an important development tool for new training programs, and can be especially useful in doing ancillary market research with customer training programs. For instance, customers can be encouraged to comment on features they'd like to see in new models or applications they're considering for your products. Often, people are more likely to respond on the spot as the thought comes to mind than having to wait for the end of a program and searching for some means of communicating with the information developer.

The authoring system, I.De. A. S. (Omnicom Associates) supports this user comment function. Whenever a user presses F5, he or she can leave a short comment which is then recorded in that user's response file. This feature has been used in beta-testing programs since users can easily note problems or comments.

LEARNING STYLE DIAGNOSIS AND BRANCHING TECHNIQUE

Purpose

Interactive systems can be programmed to infer patterns from learner responses which can aid in developing a strategy for branching. Instead of merely branching based on responses to each question, a system can keep track of response styles and cumulative scores. After sophisticated analysis, the system can access program segments and present material that corresponds to a learner's style and overall performance.

Description

Much research has been done over the past 20 years on learning style: the theory is that people have different ways of learning—some by doing, some by reading, some by listening; some are field-dependent, others are self-motivated, some are deductive, and the list goes on. Whatever your typology of learning styles, an interactive program can probably be designed to assess trainees' patterns. This is a significant goal in and of itself, and that knowledge of learning style can be useful for the trainee and the trainer in terms of future instruction, no matter what the medium. However, the concept can be taken one step further and one can have the interactive program "figure out" the learning style and present branches based on its diagnosis.

The first and most difficult step in applying this technique is to determine the categories of learning style with which you wish to deal, and then find some valid and reliable method of assessing those styles. If one is not an educational psychologist and/or isn't paid to conduct research, it is probably better to find and use someone else's instrument. Another practical way of implementing the idea is to work with only two or three basic ways of presenting content, and to try to figure out directly or indirectly which is best for the particular topic that is covered rather than trying to figure out one's overall learning style.

For instance, a program might simply offer trainees a choice among several ways of introducing a topic: examples, theories, formulas, written descriptions, or graphics could be included as branches. Content can be presented in one or two of these ways in the beginning of a program; and by asking questions, the program can determine whether, for instance, trainees tended to do better on content that was presented using simplified graphics or content presented by verbal descriptions. Of course, this technique means constructing many branches for not only right and wrong answers, but also for different methods of presentation as well. It also implies more sophisticated response analysis than is available with many interactive devices and authoring systems.

Although it offers powerful benefits, this technique must be used with caution because of the research and amount of production necessary to pull it off. It can be abused by those who don't understand instructional design and psychological testing. One common misapplication is in "tracking" in which, early in the program, learners are judged as "smart," "average," or "dumb" (actually, most designers use euphemistic terms, but it boils down to the same idea). Then, individuals are presented with branches that are appropriate for their supposed intellectual level. Unfortunately, trainees' intelligence is usually confused with their knowledge of the subject matter, and inappropriate treatments of the content are prescribed. For instance, if you were to present a program on automobile maintenance to a group of highly educated and intelligent women, you might find that few had much prior knowledge of the subject matter, so would score low on early questions. However, that doesn't mean that subsequent material should be presented in a simpler manner; it's just that more prerequisite explanation is probably necessary. Some interactive programs, though, take such preliminary information about response scores and decide that the viewer needs a slower, less sophisticated approach, and perhaps one that doesn't rely on verbalization and high-level concepts. In these cases, the audience is badly stereotyped and will undoubtedly be turned off.

Example

Learning style diagnosis is one of the most sophisticated techniques possible in interactive video, and few if any actual training courses apply it correctly. However, it is being done with computer based-instruction, and pilot projects using interactive video are underway. It is especially useful when dealing with heterogeneous groups of trainees and with content that can sensibly be presented using different styles.

In an early version of learning style diagnosis, a government agency has identified learners as either conceptual or practical. At several points in a program on resource and management, the video program stops and learners are given a choice of (1) a mathematical model for allocating resources or (2) a combined graphic and visual depiction of the actual resources being allocated to various delivery points. The program presumes that learners may not make the correct choice and offers the option to view both versions of the resource allocation segment.

VIDEOTAPE RESPONSES TECHNIQUE

Purpose

Using videotape-based interactive systems, user responses can be taped so that they and their trainers can compare their actual vocal and motor responses to models on the tape. This is an extension of the language-lab approach in which students' voices were recorded on a tape in response to a master-teacher's cues. This allows the interactive system to provide a much broader means for evaluating learning, especially in the areas of motor performance and speech.

Description

This technique requires a computer-based interactive videotape system or tape/disc "hybrid" with an interface device that is capable of putting the video machine in the "record" mode. (Of the many interactive video interfaces on the market, few offer this option.) Instructional and question segments are created as in other interactive programs, but instead of having the learners merely type in answers, they can be instructed to reply by speaking or performing some other appropriate action. Blank segments are left on the videotape, to be recorded upon by a camera and microphone hooked up to the system. At the appropriate time after the question is asked, the recorder goes into the "record mode" and the trainee must respond in a given amount of time. At the end of the blank segment, the VCR stops, disengages the recording function and the system goes back to presenting information.

Obviously, this can only be done with a few systems and it requires careful planning of a program. Many copies of the original instructional program may be needed since the tape gets filled up with student responses. Since the system includes a camera and microphone, the delivery hardware is quite expensive and adds components which can fail or become disconnected. Also, this system obviously can't evaluate learner responses by itself; it relies on trainee self-evaluation by showing them their responses in comparison to model responses, and upon subsequent trainer evaluation. Therefore, it's not appropriate for situations in which immediate feedback is needed for individuals who may not be capable of evaluating themselves.

Examples

A typical application of this approach is a substitute for role-playing. Troy State University in Alabama has produced such a program to teach and test social work skills. This technique is also being used to teach sign language, much in the manner of a video language lab. It's also being used as an assessment device for testing out of required introductory speech courses at a community college. Students can go into a private area and demonstrate their mastery of communication skills by verbally analyzing scenes, and engaging in mock conversation with individuals presented on tape. Faculty members then view the videotapes and evaluate the individual students based on a checklist. This system increases the flexibility of the assessment procedure since faculty don't need to be present when students are being assessed. Faculty can also view the taped responses separately and can compare evaluations later, eliminating some of the bias that can creep in during group evaluations. A major financial corporation also uses this approach to train employees in sales skills.

CONSENSUS TECHNIQUE

Purpose

Although most interactive programs are meant for individual viewing, group response systems enable whole classes to participate in interactive training.

Description

Often it's important to be able to train a group together because of logistics. In other situations, it's desirable because there is an additional goal of "team-building" along with imparting content knowledge. Group interactive systems can allow up to 100 individuals to respond to questions and can display summaries of responses. Programs can be designed to branch to segments based on the majority vote, or can merely display the results of the tabulation, giving the leader or trainer the opportunity to select the next branch.

In team-building and organizational development activities, an important outcome is exploring individual views and coming to consensus on issues. Interactive systems allow individuals to "vote" in private by merely pressing a button on a small keypad. Their responses can remain anonymous, or the system can be programmed to record who pressed what and make that information available to the trainer. Most systems present a colorful bar graph which summarizes inputs but does not identify individuals' choices to the group.

Group response systems allow the trainer to assess the knowledge or attitudes of the group, and tailor the presentation accordingly. For mostly "live" training, this can be accomplished with just a microcomputer, monitor, and response pads. But interactive video, of course, adds audiovisual stimuli which can be useful in presenting content or choices. Several manufacturers make interactive video interfaces which include microcomputer control of tape or disc plus the provision of receiving responses from groups of response pads. Special programming aids are available to assist in writing the software to analyze these grouped responses.

Consensus branching is ideal for accurately assessing where a group stands in relation to a particular issue or specific content knowledge. Of course, just like other group training, it is based on the majority situation, so takes away some of the individualization of learning. However, the most powerful aspect is in letting individuals know how they stand in relation to the group—especially regarding values and decisions. This technique can be an effective aid in helping a group achieve consensus by conducting several votes interspersed with discussions and presentations. Since voting is anonymous and equal, people's responses are not based on social or political acceptability, and individuals are free to change their minds without being embarrassed. This technology is powerful in training teams who will work together since it reinforces mutual decision-making.

Examples

This technique would be useful for testing employee or customer responses to new products, ideas, or policies in a group. It has been used to introduce new products to sales and management personnel and to support a part of a workshop on interactive video and computer testing design. The technology allows audience members who don't know one another to disclose information about their own opinions and allows the discussion leader to check to see if participants are following the presentation.

The Warner-Qube (now called Warner Cable Communications) system in Columbus, Ohio began testing a version of the Consensus Technique during the early 70's and continued using it into the early 1980's. When cable television was young and growth projections were inflated, the Qube system was a national model for interactive programming. Cable subscribers used response pads to participate in game shows and rate new products. Test commercials were shown to subscribers whose demographics paralleled national norms for income, ethnicity, family size, etc. Subscribers rated the commercials and results were tabulated immediately and displayed on screen. The concept was to use the Consensus Technique to involve otherwise passive viewers as active participants in decision-making about products, marketing approaches and programming.

Perhaps the most successful Qube application was its broadcasting of Ohio State football games in an interactive format. By paying to see the game on Warner's premium channel, subscribers automatically gained access to an "electronic football party." Opportunities to interact (e.g., answer questions, make decisions) not only generated consensus building among viewers, but also earned prizes for subscribers who won interactive contests. Here was a case where local interest in Ohio State football was so pervasive that a large subscriber pool participated in the interactive format. Unfortunately, expanded network coverage of Ohio State football, at no charge, forced Warner to discontinue the service.

The Qube System was ahead of its time in concept, but had difficulty involving adequate numbers of subscribers for each category of interactive programming. This practical consideration is necessary in any setting where consensus building is attempted. Adequate sample size and a representative group of participants are both prerequisites for the Consensus Technique. Limited participation reduces the value of the exercise and, in the case of Qube, the economic viability of the project.

HIDDEN PROGRAMMING TECHNIQUE

Purpose

This technique allows you to imbed information on the audio, video, or computer segments of a program that is meant for special audiences or situations only. By doing so, you can make one program serve several purposes, or include special instructions about administering the training right in the software.

Description

"Hidden programming" sounds much more mysterious than it really is. All it takes to create this effect is to have a branch or so which has no obvious means of access–unless the user learns the "secret" accessing code in some way. Let's say that an organizations wants to create a program on cafeteria sanitation for food handlers–but also wants to include some special information for their managers. Perhaps the main menu states three options for the food handler trainees, but, in a special memo to managers included with the tape, they're told that they should push "4" when they see that menu. This response, known only to them, would access one or several branches meant for managers while allowing them to see whatever other parts of the program are appropriate.

The main problem in creating this effect is selectively informing individuals of the "secret" code to get into their portion of the program. Of course, with microcomputer-based programs, sections can be "unlocked" based on ID numbers or passwords. Some systems begin by asking what segment the user would like to start with; here, one can inform managers that for an entire series of training programs, they should always press "99" and course administrators should always press "007" or some such catchy number. The branches can of course be separate segments, or can merely access a different audio channel with a separate voice-over tailored for the special audience.

Besides screening individuals from certain information, secret programming codes can also be given to individuals when they complete a course allowing them to proceed to another course or teaching them how to skip around and review the course they just completed. This gives trainees flexibility in reviewing materials and prevents them from randomly skipping around in and among programs before they're ready.

"Hidden programming" is found in many programs ostensibly meant for customer information and training; there are often separate segments there which only company employees or dealers can access. Often, this material includes sales strategies, comparative sales information, or pricing structures. Several automobile videodiscs as well as some videodiscs about interactive video have such segments.

This technique must be handled, however, with sensitivity. If the primary audience finds out that there are segments which they can't see, they'll probably be annoyed and offended (they'll also probably find a way to see it). Certainly, highly confidential information should not be placed in such segments since it's an easy matter to override the computer control and play back the whole tape or disc, or merely run a copy of the computer diskette.

Example

Hardinge Brothers uses a videodisc/touchscreen simulator to train customers to program and set up their computer-controlled lathes. Although it is undesirable to allow trainees to randomly "skip around" in the program (especially to avoid difficult sequences) it is necessary to provide the trainer with that capability. There are areas of the screen (known only to the training staff) which can be pressed to branch the user to the main menu, or to special sections for one-one-one tutoring session with the trainer.

INTELLIGENT "RECURSIVE" PROGRAMMING TECHNIQUE

Purpose

Mediated systems, up until now, could only do what they were told to do—they could only present what the course author had specifically created. New advances in intelligent systems enable trainers to create recursive programs which modify themselves *during* use and can actually "learn" as they teach.

Description

A problem with many interactive systems is that they can't accept a wide enough range of responses and deal "intelligently" with them. For instance, if a trainee uses a perfectly acceptable synonym for a term, but that synonym wasn't anticipated and programmed in by the course author, the system will either: (1) consider the response incorrect, (2) lump the responses in with all unanticipated responses and present a generic "huh?" branch, or (3) hang up or present a continuous boring loop until the trainee figures out some other word to use. Recursive systems allow the program to query trainees when they use unanticipated vocabulary; if the student can explain the concept in other terms that the program *does* "understand," it will add their new word to its vocabulary list. Just like a good teacher, the program becomes more sensitive to trainee language and applications through practice.

Recursiveness has been demonstrated in computer-based-training and expert systems, but is not widely adopted in interactive video—although there's no real barrier to doing so. Intelligent systems do require special high-level programming, though, and must be monitored carefully in use. For instance, it's possible for trainees to inadvertently or deliberately "teach" the system something wrong (just like humans can sometimes be fooled), so it's important for a trainer to periodically examine the program to see what it's added to its vocabulary list.

Self-modifying programs have traditionally been executed using list-oriented programming languages like LISP and FORTH, however, LOGO, a popular language for teaching children about computers, is a recursive language, and other intelligent systems have been programmed in BASIC. The major programming problem here is to write a program which can create a new version of itself—to write a program which creates a program. If you think it's difficult to create media that teach humans, try creating a program that teaches and learns!

Examples

In situations in which you want to teach specific terminology, recursive systems would not be applicable. However, if you're not exactly sure of what wide range of wording trainees might use to respond to questions, you might want to include recursive sections in an interactive program. This technique is very effective for systems that will be used by the general public to provide broad-based information, since the users will probably employ varied vocabularies in interacting with the system, and it is difficult for the program author to anticipate all of their terminology in advance. The recursive system will gradually learn more vocabulary to deal with this variety of responses, and learn how to deliver appropriate feedback. This kind of recursive system automatically modifies itself so that trainers and producers can see what exactly users have been "saying" to the program. This kind of record-keeping is as if the course developer could always be leaning over the shoulder of users, and modifying the program based on their experiences with it. Although this "artificial intelligence" approach to design is not applied broadly as of the writing of this book, it is certainly a technique under development by software engineers, and should surface in many training and information applications fairly soon.

Bibliography

Brush, Judith M. & Douglas P. Brush. *Private Television Communications: The New Directions (The Fourth Brush Report).* Cold Spring, NY: HI Press of Cold Spring, Inc., 1986.

DeBloois, Michael L. (Ed.) *Videodisc-Microcomputer Courseware Design.* Englewood Cliffs, NJ: Educational Technology Publications, 1972.

Fleming, Malcolm, and Levie, W. Howard. *Instructional Message Design.* Englewood Cliffs, NJ: Educational Technology Publications, 1978.

Floyd, Steve, and Floyd, Beth. *Handbook of Interactive Video.* White Plains, NY: Knowledge Industry Publications, 1979.

Gayeski, Diane M. *Corporate and Instructional Video: Design and Production.* Englewood Cliffs, NJ: Prentice-Hall, 1983.

Gayeski, Diane M. *Interactive Toolkit.* Ithaca, NY: OmniCom Associates, 1987.

Gayeski, Diane M., and Williams, David V. *Interactive Media.* Englewood Cliffs, NJ: Prentice Hall, 1985.

Hilliard, Robert L. *Writing for Television and Radio* (3rd ed.) New York: Hastings House Publishers, 1976.

Iuppa, Nicholas V. and Anderson, Karl. *Advanced Interactive Video Design.* White Plains, NY: Knowledge Industry Publications, 1988.

Kemp, Jerrold E. *Planning and Producing Audiovisual Materials* (4th ed.) New York: Harper & Row Publishers, 1980.

Matrazzo, Donna. *The Corporate Scriptwriting Book.* Portland, OR: Media Concepts Press, 1980.

Millerson, Gerald. *The Technique of Television Production* (9th edition). New York: Hastings House Publishers, 1972.

Schwier, Richard. *Interactive Video.* Englewood Cliffs, NJ: Educational Technology Publications, 1987.

Utz, Peter. *Today's Video.* Englewood Cliffs, NJ: Prentice-Hall, 1987.

Index

Alter ego technique, 54-58
    guidelines for, 50-51
Animation and graphics techniques, 77-79
Audio. *See* Sound
Audio lead technique, 21-25
Auditory/visual test technique, 150-152
Authoring systems, 105

Branching, 102, 104
    in response to learning style, 160-162

Camera focus, 84
Camera lens and depth of field, 83
Camera movement, 84
Character perspective, 83
Character tracking technique, 49-53
Commercials
    compared with linear designs, 13-14
    using linear design techniques, 14
Consensus technique, 165-167
Controlled playback technique, 111-112
Cue fading technique, 80-81

Dedicated microprocessors, 104
Deliberate degradation of video and/or audio technique, 88-89
Deliberate overstatement (exaggeration) technique, 90-97
Digital squeezing transition, 62
Direct address approach, 102
Directed simulation, third person, 135-136
Dramatic irony technique, 97-100

Eighteenth hole voice technique, 15-20
Exaggerated sound, 86
Exaggeration technique, 90-97
Examples, relevant, 127-130

Fast motion, 86
First person simulation technique, 137-139
Flashforward technique, 72-76
Freeze frame, 86

Gaming technique, 142-144
Graphic replacement, 86
Graphics and animation techniques, 77-79
Group training
    consensus technique for, 165-167

Hidden programming technique, 168-169
Humor, importance of, 12-13

Intelligent manual technique, 117-120
Intelligent "recursive" programming technique, 170-171

Interactive design techniques, 101-171
- advantages of, 106
- anticipating responses in designing, 106-107
- auditory/visual test, 150-152
- consensus, 165-167
- controlled playback, 111-112
- direct address approach, 102
- equivalent to "Level II" videodisc, 104
- first person simulation, 137-139
- gaming, 142-144
- hidden programming, 168-169
- intelligent manual, 117-120
- intelligent "recursive" programming, 170-171
- introductory menu, 113-116
- learning style diagnosis and branching, 160-162
- levels of interactivity in, 102-106
- opinion question, 148-149
- pause, 102, 108-110
- pretest, 124-126
- recursive, 106, 170-171
- requiring constructed answers, 105
- response peripheral, 105, 153-157
- selecting relevant examples, 127-130
- simulation with feedback, 140-141
- third person "directed" simulation, 135-136
- tutorial, 145-147
- user comment file, 158-159
- using dedicated microprocessors, 104
- using microcomputers interfacing with tape or disc, 104-105
- using random-access controllers, 102, 104
- using simple branching, 102, 104
- vicarious travel, 131-134
- videotape responses, 163-164
- visual data base, 121-123

Interactive video, 104-105
Introductory menu technique, 113-118

Learning style diagnosis and branching technique, 160-162
Level II Systems, 104
Level III Systems, 105
Linear design techniques
- alter ego, 54-58
- audio lead, 21-25
- character tracking, 49-53
- compared with commercials, 13-14
- cue fading, 80-81
- deliberate degradation of video and/or audio, 88-89
- deliberate overstatement (exaggeration), 90-97
- dramatic irony, 97-100
- eighteenth hole voice, 15-20
- flashforward, 72-76
- graphics and animation, 77-79
- humor enhancing, 12-13
- importance of audio to, 10-11
- off-screen audio montage, 26-30
- omniscient spokesperson, 33-38

out-of-character commentary, 39-42
parallel scenes (action), 59-65
quick visual inserts, 65-71
repetition, 82-87
scriptwriting tools for, 9-10
subjective viewpoint, 43-48
used in commercials, 14
use of talent in, 11
using music, 31-32
visual variations to improve, 11-12

Music, using, 31-32

No sound, 85

Off-screen audio montage technique, 26-30
Omniscient spokesperson technique, 33-38
Opinion question technique, 148-149
Out-of-character commentary technique, 29-42

Parallel scenes (action) technique, 59-64
transition options for, 60-62
Pause, 102
Pause technique, 108-110
Pretest technique, 124-126

Quick visual inserts technique, 65-71

Rapid dissolve transition, 61
Rapid fade transition, 61
Recursive programming, intelligent, 170-171
Recursive systems, 106
Relevant examples, 127-130
Repetition technique, 82-87
production vehicles for, 83-86
Replacement sound, 85-86
Response peripherals, 105
Response peripheral technique, 153-157
Reverse motion, 86

Selected sound, 85
Selecting relevant examples technique, 127-130
Simulation
with feedback, 140-141
first person, 137-139
third person "directed," 135-137
Simulation with feedback technique, 140-141
Slow motion, 86
Sound
audio lead technique, 21-25
auditory/visual test technique, 150-152
deliberate degradation of, 89
eighteenth hole voice technique, 15-20
exaggerated, 86
importance of, 10-11
manipulation of, 85-86
no sound, 85
off-screen montage technique, 26-30
omniscient spokesperson technique, 33-38
out-of-character commentary technique, 39-42
replacement, 85-86
selected, 85

using music technique, 31-33
Sound manipulation, 85-86
Split screen transition, 61
Straight cut transition, 60-61
Subjective viewpoint technique, 43-48

Talent, importance of, 11
Third person "directed" simulation technique, 135-136
Timing effects, 86
Tutorial technique, 145-147

User comment file technique, 158-159
Using music technique, 31-32

Vicarious travel technique, 131-134
Video
audience as viewer-performers, 8
decisions to make before using, 5-6
importance of audience involvement with, 3-4
interactive design techniques for
See Interactive design techniques
judging effectiveness of, 3
linear design techniques for
See Linear design techniques
need to know audience for, 8-9
task-specificity important to, 8
training use of, 4
viewer aptitudes important to, 8-9
Videodisc/computer systems, 105
Videotape responses technique, 163-164
Viewer aptitudes
importance of, 8-9
Visual data base technique, 121-123
Visual techniques, 12
Visual variations
importance of, 11-12
Voice-over, 15-20

Whip pan transition, 62
Wipe transition, 61-62